Smarter than Joey

Smarter than Joey

91 true animal stories from Australia

Avocado Press Limited

Published by:
Avocado Press Limited
PO Box 27003
Wellington
New Zealand
info@avocadopress.com
www.smarterthanjack.com

Cover design: Simon Cosgrove
Typesetting: Dominic Hurley
Cover photograph: Jenny Campbell
Cover subject: Wallaby at Staglands Wildlife Reserve, New Zealand
Story typing: Gemma Sare
Proof-reading: Vicki Andrews
Prepress: Megalith
Printer: Printlink
Accounting software: Fastbase Development Company
Compiler: Jenny Campbell

Unattributed passages are by Jenny Campbell

Printed in New Zealand
First published 2004
Copyright © 2004 Avocado Press Limited
ISBN 0-9582457-1-1

All rights reserved. This book may not be reproduced in whole or in part, nor may any part of this book be reproduced, stored in a retrieval system, or transmitted in any form or by any means, electronic or mechanical, including photocopying, without written permission from the publisher, except for the inclusion of brief passages in a review. The views expressed in this book are not necessarily those of the Royal New Zealand SPCA, RSPCA Australia, or Avocado Press.

Contents

Introduction	vii
Acknowledgements	viii
Foreword	ix
1 Smart animals know what's best	1
2 Smart animals take control and outwit us	9
3 Smart animals to the rescue	19
4 Smart animals do strange things	37
5 Smart animals cope with life and death	49
6 Smart animals think of their tummies	63
7 Smart animals help other animals	77
8 Smart animals show insight	89
9 Smart animals make us wonder	99
10 Smart animals help us out	123
Afterword	143
Future books you can be involved in	144
Submit a smart animal story	145
How to get more smart animal books	147
We'd like to hear your ideas	149

Nelson, or is that Lassie? Read his story on page 35.

Introduction

Animals are fascinating, especially when they catch us out with their intelligent antics. Perhaps you've seen kangaroos enjoy a game of soccer, or a horse fetch the cows in for milking – all by itself! In this book you'll read these and many other true stories about smart animals. You'll start to wonder what's really going on inside their heads.

In 2003 people throughout Australia submitted several hundred true stories. Ninety-one of the best were selected for the Australian edition of *Smarter than Jack* and so far over A$45,000 has been raised for RSPCA Australia.

Smarter than Joey is a special edition with these Australian stories for New Zealanders to enjoy. Profit from sales will help the Royal New Zealand SPCA.

There are stories about kangaroos, cockatiels, crows, bandicoots, wombats, possums, galahs, dogs, cats, horses, skinks, chooks, spiders, magpies, sheep, dolphins and a fox.

In New Zealand two books have been published in the *Smarter than Jack* series, both were bestsellers. The first, *Smarter than Jack*, was published in 2002, raised over $43,000 for the Royal New Zealand SPCA and won the Bookseller's Choice Special Recognition Award. *Smarter than Jill* followed in 2003 and has raised over $40,000 to help animals. Both books remain very popular.

The future of the *Smarter than Jack* series holds a number of exciting books. We'd like to get your stories now. See the back of this book for details of how to be involved, or visit www.smarterthanjack.com.

We hope that you enjoy *Smarter than Joey*. It has been a very fulfilling book to create and we hope that many people and animals continue to benefit from it.

Acknowledgements

This book really belongs to everyone who has known or read about a smart animal. I find it strange putting my name on the cover, because so many others have had a hand in its creation.

This includes everyone who submitted a story, and especially those who had a story selected as this provided the content for this book. The people who gave us constructive feedback on earlier books and cover design, and those who participated in focus groups, helped us make this book even better.

The teams at RNZSPCA and RSPCA Australia and their member societies were keen to be involved, and assisted us greatly. They helped to get the need for stories and the book known.

Anton, my partner, gave me continued support and put up with my often obsessive work hours.

Kerri-Anne Kennerley wrote the wonderful foreword, Simon Cosgrove designed the cover, Gemma Sare helped type the stories, Vicki Andrews did the proof-reading and Dominic Hurley did the typesetting.

Production partners Megalith and Printlink did a grand job of producing the book.

Lastly, I cannot forget my endearing companion, Ford the cat. Ford has not yet enthralled with his superior intelligence, but that could simply be because I'm such an inferior being to him and his intelligence is way beyond my level of understanding. After all, who is the one who feeds him when he meows, pets him when he jumps on my lap and opens doors for him on demand!

Jenny Campbell
Avocado Press

Foreword

When I read through the amazing stories in this book I was reminded of a very special friend. The following tribute is what I wrote when I lost Murphy . . .

Murphy had one ear, one eye and one ball and we called him Lucky . . . not really. He did lose an eye and an ear canal and had a masculine bit removed, but his name was Murphy – a spritely brown and white springer spaniel. Murphy was a survivor and a fighter – a bit like his mother, I was told. An independent thinker, our Murph was well behaved and disciplined, but stubborn as anything. It was always more successful to ask him nicely.

For many years, I rose at 4 am every day to trek off to the TV studio. It can be a lonely hour and it's easy to disturb a partner's sleep, so I would creep out of bed as quietly as possible and tiptoe around with my loyal boy at my side. Murphy would unfailingly follow me to the bathroom, lie there while I did my make-up and dressed, and then wag me goodbye at the dark front door. Murphy loved his home. He loved going out, but when he'd had enough he would stand stubbornly and demand to be taken back to his own environment.

Occasionally, I would amuse myself and break out the karaoke for a bit of private fun, and Murphy would sit looking on intently as I sang at the top of my voice. I realised after a few sessions that he was pretty well deaf, but for a while there I thought he really liked my singing. He didn't like change a lot and would follow me close at heel, sometimes driving me mad. Oh, how I'd love to be driven mad right now – just to turn around and trip over him.

Murphy's arthritis had been playing up with the cold weather, so it was time to review his medication. Off to the best and most caring vet in the world, Dr Rob Zammit. Rob had said after a few days that

Murphy was doing very well and, as I was off to Perth to work that weekend, Rob suggested that I leave Murphy with him.

He was a 14-year-old dog, and took a turn for the worse. It happened out of the blue and nothing could be done. Rob had tests done to reveal that he was suffering kidney failure and, as I couldn't bear to have him in pain a second longer than necessary, with Rob's advice and tender loving care Murphy was at peace. I was at least comforted that he was in Rob's care. But I wish I could have kissed his freckled face, scratched him behind his good ear – which always made him moan in ecstasy – and seen a last wag of a stubby tail. I wish I could have held him to let him feel how much John and I appreciated his lifelong dedication to us. I couldn't say goodbye, Murphy, so this tribute is the least I could do.

Well, years later I still miss him but I have a portrait to remind me of him. Nothing can ever replace a friend – a pet with whom you shared time, love and devotion – but I would urge anyone who has been saddened by a loss to move on with a new housemate as John and I have. Harvey – named after my dear friend Geoff Harvey – has fitted in brilliantly. In fact, he has certainly taken over in many respects. He didn't grow up with a doggy friend as Murphy did with our retriever Angus. So, being an only dog, he took control. We love him as much, but differently. I only wish everyone could know the same affection.

Kerri-Anne Kennerley

1

Smart animals know what's best

He knew what to do

Sandy was a lovely old yellow Lab who lived till he was 14. When he was about ten years old we took him to the vet (a few blocks away from where we lived) with a wound in his side.

The vet dressed the wound and put a number of stitches in, and said to bring him back in two weeks time to get the stitches removed. I checked the wound from time to time and everything was healing beautifully. By the end of the two weeks the stitches had just disappeared so I did not bother to go back to the vet.

Three months after this, Sandy was due for his annual vaccination. Feeling a bit embarrassed about not going back to get the stitches removed, I told the vet why I had not returned.

He said that Sandy had come round every day for treatment, waiting at the back door for him to clean the wound. On the final day, when the vet removed the stitches, he told Sandy that he did not have to come back any more. Sandy never went back until the day we took him.

Graham Evans
Bullsbrook
Western Australia

Not to be underestimated

I know animals are smart. I've been around animals a long time and it doesn't take all that long to learn that most animals have minds of their own.

One day I was sitting on my kitchen porch and I heard a rustling noise down in the garden below. I thought it might be a snake as we had many of them in the country, and I wanted to make sure I knew where it went. There did not seem to be a snake, but the rustling noise continued and I thought I had better go downstairs to have a look at what was happening in the garden.

Much to my surprise, I found a skink tangled in some wire netting that I had put around a small tree to prevent the dog from scratching on it. As it struggled to free itself I wondered how I could help without hurting the lizard.

I didn't need to think very long, because suddenly another, bigger skink appeared on the scene. I retreated and watched from a distance where I would not be a nuisance to them. The bigger skink walked around the one caught in the wire as if summing up the scene before him. Then he pushed against the smaller skink with all his weight, as if to push him through the wire. This didn't work, the smaller lizard was still stuck.

Then the bigger lizard set to work with his teeth. He grabbed the wire and pulled it this way and that, trying to loosen it from where it held the other lizard so firmly. I was amazed. I had never seen this sort of thing before. I had never thought that lizards had any idea of much. After all, they are such quiet, unassuming little creatures.

I watched as the big skink continued to tug against the wire, moving up and down the other skink's body until he had lessened the strain of the wire. Then he went to where most of the other skink's body was pressing through the wire and gave a great shove with his body. All of a sudden the smaller skink was free. I cheered. I was absolutely delighted at what I had seen, and watched happily as the two lizards ran off together into the bushes.

I feel I was honoured that day to see something that not many people have seen, two skinks interacting with each other in a very human way.

Bernie Pattison
Redcliffe
Queensland

Just so

Having moved to a 50-acre block in Longford, near Sale, Victoria, we acquired some beautiful hens.

Visiting the large chook shed, we noticed that the depression in the hay in one of the nesting boxes was in the corner of the box. We thought this must have been uncomfortable for the chooks so my husband carefully rearranged the hay, placing the depression in the middle, which we thought would be more convenient for the hens.

Just as the task was completed, one hen entered the shed and inspected the box. She clucked, and clucked again, then departed. She returned with Rudy the rooster and the rest of the girls. They all looked in the box and clucked like a bunch of old women.

The hen then jumped into the box and rearranged the hay so that the depression was back in the corner.

I would not have believed this had I not seen it for myself.

Betty Tanava
Pomona
Queensland

A hard lesson

Before immigrating to Australia, my late husband and I lived in Umhlanga Rocks near Durban in South Africa (although we were both born in Manchester and lived in the United Kingdom).

One day, we were driving along the busy main road towards Durban when we saw a very large monkey and her baby sitting alongside the road. Next, without warning, the baby dodged through the traffic to the opposite side of the road, landing there unharmed.

Equally suddenly, the mother darted across and on reaching the child boxed its ears in no uncertain manner, just as a human would do with a naughty offspring.

I don't think that baby would attempt such a manoeuvre again.

Evie Perrins
Applecross
Western Australia

In a jam

This story was told to me by Alex McDonell of Kunderang Station on the Upper Macleay. He had a cattle bitch that had puppies underneath the house – the working dogs all lived outside and never came into the house.

One morning he heard one of the puppies whimpering in obvious distress, but as it was completely inaccessible there was nothing he could do to help it. When he came home that night the bitch brought the puppy to him in the kitchen. In those days, jam came in tins, and when the tin was finished the top was pushed in and the tin discarded. The puppy had put its head into the tin to lick out any remaining jam and could not get it out again, being trapped by the pushed-in top. The bitch, knowing she could do nothing to help the pup, brought it in to Alex, who released it and gave it back to its mother. She took it back under the house, never coming inside again.

Alex told me the story to illustrate the intelligence of the dog, but it also says a lot for the trust she showed in her gentle master.

Alex told me of another dog that belonged to his neighbour Laurie O'Keefe of Oatlands, Yarrowirch. They were on a droving trip when one morning Laurie said to the dog, 'Bring them in.' The dog ran off and started to round up the cattle, but Laurie called out, 'No, not them, the horses.' The dog then left the cattle and ran the horses into the yards.

> Anne Holberton
> Willawarrin
> New South Wales

Right to the source

My dog Barrie, a Rottweiler, adopted our family a number of years ago. Shortly after he joined us he showed fright at the noise of the vacuum cleaner. He went straight to the power point and removed the plug!

> Monica Steven
> Trinity
> South Australia

Tonia's trip to the vet

I had a beautiful big black Newfoundland called Tonia. We had a ritual for our daily walk: when we got to the gate, Tonia would choose which way we would go. She knew the routes well enough after all those years to know what going to the right as opposed to the left would offer.

This particular day she chose to go left. She didn't choose this way very often because it was a shorter walk. I hesitated, but she pulled me quite firmly in that direction as only a 58 kilogram dog could! There was a real purpose in her stride this day. Usually she ambled along at her own pace because now she was getting old.

We got to the end of the street, turned left and headed up to the main road. Tonia usually turned right there. We rarely walked this way because of all the traffic. It meant we had to wait for a break in it and dash across the road because there were no lights or crossing. I was becoming more curious by the minute as I let her lead the way.

We approached the roundabout and negotiated the traffic there, crossing safely. Instead of heading to the walking track, we continued along the main road. Then it clicked. This was the way we walked to the vet sometimes. Surely she wasn't taking me there! Tonia was no different to most animals, she didn't particularly like going to the vet. Why would she head this way if that wasn't the reason?

We approached the driveway we used to cross the road to the vet. This was the moment of truth. Sure enough, she went down it and sat waiting for the traffic to clear before we crossed. Tonia trotted quite purposefully through the automatic glass doors and parked her bottom right in front of the reception desk, tongue lolling and looked up at me expectantly.

The receptionist greeted us, assuming we had an appointment. My jaw had dropped by now and I was speechless. Gathering my breath, I explained the story to her. I asked if we could see a vet since Tonia had gone to all the trouble of bringing herself there. I was expecting a long wait because we hadn't made an appointment.

Just then, the vet we usually saw poked his head out of the consulting room and said hello. Still in a state of amazement I told him the story. As it happened he was free and could see her right away. On examining her, he discovered she had a urinary tract infection that required treatment! Tonia must have had enough of the discomfort from the infection so had taken action as I obviously hadn't picked up that she was unwell.

On the way home she reverted to her usual ambling pace, and I held a white paper bag containing her antibiotics. I couldn't wait to get back to tell my husband the story. He was as amazed as the receptionist, the vet and I were. Who says animals aren't smart!

Julie Biro
Mt Evelyn
Victoria

2

Smart animals take control and outwit us

Magpie discipline

I rested on a log one afternoon to watch two magpies feeding, a mother and her well-developed youngster. Mother trotted about her patch collecting grubs, worms and other tasty morsels for her insatiable offspring.

The pressure of maintaining a steady stream of food for junior apparently became too much for her. The next time the youngster opened his mouth and demanded food he received a compact wad of dry grass. It took him several minutes to clear the grass from his beak and throat.

Mother enjoyed her free period by swallowing as many grubs as she could find in the time available to her. She then resumed feeding her rather surly youngster.

> Frederick Becker
> Nambour
> Queensland

Like my limp?

The following is a true story about my Belgian Shepherd called Blaec, who I think is a very smart boy. He constantly amazes me with his almost-human antics . . .

At the time this took place, Blaec was about two years old. My cat Milo ran away and was missing for two weeks. When he came home we saw that he had been caught in a fox trap and had injured his front right paw. After a trip to the vet and an operation, Milo's leg was in a cast and he was forced to stay indoors. If he wanted to go outside we had to carry him, as he wasn't allowed to get his cast dirty.

Blaec soon became jealous of all the attention Milo was getting – he started sulking and doing almost anything to draw attention to himself.

About two days after Milo's trip to the vet, Blaec suddenly developed a limp, refusing to put weight on his front right paw. I checked his paw to make sure there were no prickles in his pads or between his toes. I even twisted his leg round to check for strains, but couldn't find anything wrong with him. However, when he stood up he was still limping badly.

I was struck with an idea, so I went inside the house and got some sports strapping tape. I stuck it around Blaec's leg like a bandage. I then made a big fuss of his sore leg. Miraculously, he made an almost instant recovery – the limp was gone and he was walking normally again – he was too distracted by all the attention to remember to limp!

Sarah Jones
Conder
ACT

Lateral thinking!

Whenever leaving our house for the day we encouraged our tortoiseshell cat Tammy to spend her time in the rear garden, weather permitting. She usually required persuading from hiding in the family-dining area of the house.

One day we failed to find her in the family-dining area despite searching in the usual hiding places. We decided to do another search, being convinced that she was there but somehow managing to elude us.

The family room has two recliner rocker chairs with covered bottoms, which were Tammy's occasional hiding place. On our first search we tilted the rocker chairs forward as usual but found nothing underneath. On the second search, tilting one chair higher and looking more closely, we found Tammy stretched between the two rocker feet, which allowed her to move upward with the chair without touching the floor and being seen.

We were amazed by her lateral thinking. To our disappointment she never repeated the manoeuvre.

Ludwig and Lyn Komorowski
Applecross
Western Australia

Doggy dumb-bell

A friend had bought Clancy and Keira a doggy dumb-bell for their Christmas present. The dumb-bell, when they carried it around, would make different noises. Both dogs loved it very much.

Clancy is usually a very good-natured dog and is happy to share his toys and food with Keira. That all changed when the doggy dumb-bell was presented to them on Christmas Day. There would be no sharing of this toy, and Clancy claimed it for his own. If Keira even went close to the toy he would immediately snatch it up, snarl at her, then put it next to him or sleep on it. If Keira should be lucky enough to get hold of the dumb-bell, Clancy would chase her until she dropped it or he would take it directly out of her mouth.

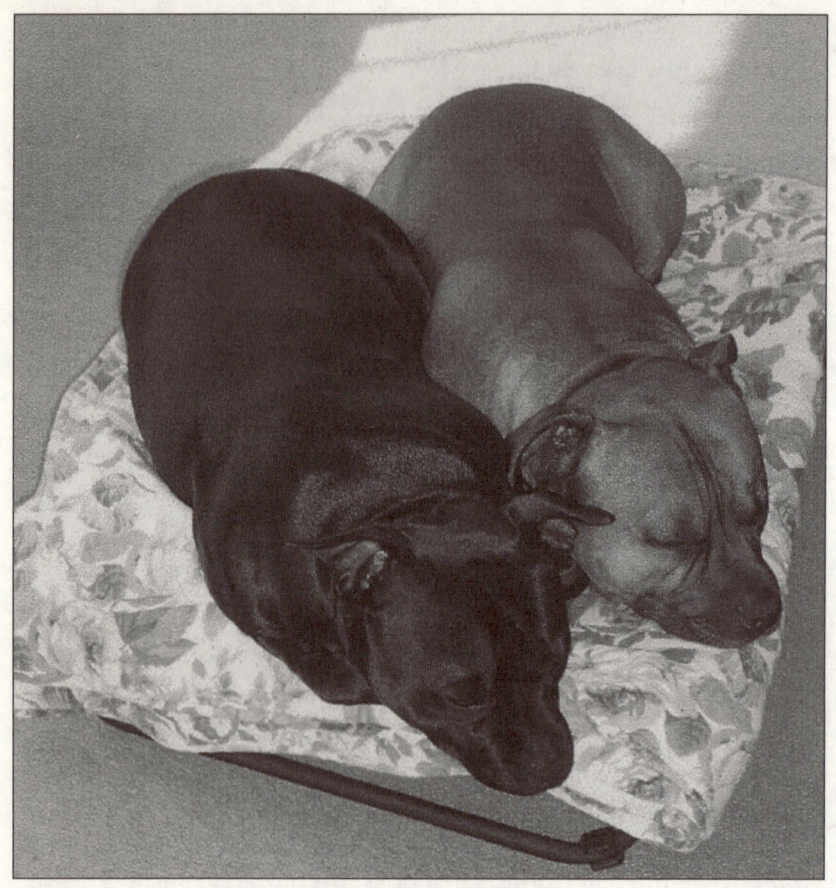

Clancy and Keira asleep

This particular day, Clancy was sleeping in the lounge with his head resting on the bar of the dumb-bell and Keira was asleep on the other side of the lounge. Clancy woke up and needed to answer the call of nature. He got up off the couch and, after checking that Keira was still asleep, proceeded to go outside to do his business. As soon as Keira heard the dog door flap indicating that Clancy was outside, she ran over to the dumb-bell, grabbed it, ran into our bedroom and put it under the bed, then ran back to her spot on the couch, lay back down and pretended to be asleep.

Clancy came back inside and straight away noticed the dumb-bell was missing. He went to Keira and started pushing at her with his nose; she still pretended to be asleep! He pushed her a bit more with his nose and was trying to look underneath her when she slowly opened her eyes, stood up, gave a big stretch and looked at Clancy like she was saying *What?* Clancy then did a big search of the lounge room, including under the cushions and behind the couch. Keira quietly looked on and then proceeded to entice Clancy into a game and he forgot about the missing dumb-bell.

When we went to bed that night Keira wanted to sleep under our bed instead of on her own bed. Clancy went to his bed and proceeded to go to sleep until he heard the noise of the dumb-bell coming from under our bed. The chase was on again.

Cathy Gillot
Worrigee
New South Wales

Outsmarted by a horse!

Before weight got the better of me, I spent eight or nine years employed as a jockey. Part of a jockey's employment is horse riding work in the mornings for whichever particular stable he or she is race riding for.

Although I came across a lot of horses you could regard as characters, one in particular stands out in my mind. This horse had a reputation for clowning around. A football minus the bladder was kept in his yard. He and the horse in the yard next to him would occupy their time by throwing the ball to each other. Should the ball go over the wrong fence, he would make sure the whole neighbourhood knew about it until someone retrieved it for him.

One particular morning I was saddling up a horse in a nearby yard. One of the stable girls went into his yard with the gear to saddle him up, but forgot the bridle. She put the saddle on a rail at the back of his yard next to a super six fence and went back to the tack room to get a bridle. While she was away the horse picked the saddle blanket up in his mouth and dropped it over the fence. On her return, seeing there was no saddle blanket she went back to the tack room to get another one. As soon as she left the yard he did the same thing with the blinkers.

Word got around pretty quickly and soon most of the workers in the stable were looking on, watching this horse outsmart his strapper. He pulled the same stunt twice more that morning with two other items of horse gear.

The laughter of her co-workers gave the strapper the idea that something was going on, but she would not believe (and still doesn't) that the horse was the one hiding the gear on her.

And, by the way, the strapper involved was no dill – today, she is a highly qualified vet.

Anon

Gullible!

Once we had a white Maltese terrier called Beaulah and a young Alsatian bitch (she was at this time) called Teddy.

I would feed them outside and would put their bowls down a fair way apart. Teddy often looked at both the bowls and then would run down to the corner of the house and start barking furiously.

Beaulah would run down to see what was happening. Teddy would immediately run back and get stuck into Beaulah's food. By the time poor Beaulah got back, half her meal was inside Teddy.

Beaulah fell for it every time.

Lillian Clayton
Redcliffe
Queensland

The three Siamese who'd had enough

Thirty years ago when I was a young teenager, I was blessed with the companionship of three Siamese cats whom I had named Ting, Ming and Ling. They were different colour types – tabby point, chocolate point and lilac point. Two were distantly related. All had very different personalities. Their capers and escapades were a constant source of amusement and joy. I miss them still.

At one time in our street a new family moved in, and the two little girls often accompanied their mum down to the corner shop. Then, a little while after this, this family acquired a young and silly Alsatian and he was sent down the road with the little girls – unleashed.

Our front gates did not often get closed, so the first I knew of the Alsatian's arrival was when three flashes of white lightening streaked up the driveway and bolted up various trees at the back. The Alsatian had great fun every time it came down the street, crossing hither and thither, chasing everything it could – especially my poor cats. They did look funny – fur all puffed up, with their tails turned into bottle-brushes – but it was stressful for them, particularly for the little lilac point female.

Then one day I was standing on our enclosed front veranda when the little girls wandered down the opposite side of the street with their nuisance dog. The three cats just happened to be in the small front yard on the driveway. They were close together.

As usual, the dog rushed over but stopped dead at the driveway entrance – its front legs splayed out, head down and tongue lolling, ready to pounce. At this moment the cats, in perfect synchronisation, turned to face the dog. I couldn't believe my eyes because then, like cowboys before a gunfight, they slowly and deliberately marched down the drive, tails erect, hackles up and fur puffed out as far as possible. There they were, shoulder to shoulder in a line, in order of size from largest to smallest – Ming, Ting, then Ling. Goodness knows what their faces looked like, as I was witnessing this from above.

The dog, staying in pounce position, watched for some moments, then turned and fled! Other things occasionally frightened my cats, but never again did that Alsatian cross the street to visit us.

Susheela Millburn
Buderim
Queensland

Zac to the rescue

My little chihuahua cross Zac has a tender heart.

I bought a wind-up mouse for my cat Tigger before Christmas. The mouse ran across the floor and Tigger pounced on it, jumped on it, threw it in the air, rolled on it and bit it. Zac sat with his head on the edge of his basket, watching this awful display.

After about five minutes he could stand it no longer. He jumped out of his basket, grabbed the mouse from the 'claws of death', took it into his basket and hid it under his bedding, growling fiercely all the while.

He kept hold of that mouse for about eight months. He hid it in his bed and, when he left the bed unattended, he hid it all over the house.

However, one day he took it into the garden and hid it permanently. It was never to return. I reckon he got sick of the responsibility of care!

Anita Jaffer
Cooroy
Queensland

Popeye

We met on either side of a broom in his home – the garage of an acquaintance. He, a small bundle of ears, eyes and fleas, and me, a large two-wheeled noisy beast. He was five weeks old but only half the size of a normal kitten. His feline mother had rejected him and, despite this, or maybe because of it and the fact that his fate was imminent drowning, he settled down on the only soft surface in his home (my scarf, which had just dropped to the ground).

It was 1977 and a bitterly cold April day in Launceston. I was in my last year of college and still living at home. The family already had a cat – supposedly mine, but in reality she only liked my father – and I had been given the decree: no more cats.

So what was to be done? I knew I couldn't take all three kittens home despite the threat of drowning, but if I fronted up with one – well, I guess if the worst came to the worst I could always find a flat.

Incarcerated in an ice cream container strapped onto the tank of my motorbike, Popeye (as he was later to be named) made the four-hour freezing journey back home, miaowing pitifully all the way – well, wouldn't you?

The welcome was as I had expected.

Later, Mum stood in the middle of my bedroom in dressing gown and slippers, arms folded sternly, and asked what I was going to do with this cat – if he indeed survived the night, which she doubted. He clearly couldn't stay because I already had a cat. I blubbered, wild thoughts flitting through my mind of trying to pass a double elective in my final year as well as finding and moving into new accommodation with Popeye.

At this point, Popeye – a veteran of survival in his five short weeks of life – realised he'd hooked himself to a complete loser and he needed to take the reins. He very cutely walked up to Mum and sat on her slippers, looking up imploringly into her face. Now, Mum is a bit of a softie with animals so, steeling herself to uphold the righteous order of things, she took two paces back and repeated the question. We have to remember here Popeye's powers of survival, so, undeterred, he took 15 little steps forward, again sat on her slippers and gave her the silent miaow.

I don't remember much after that but it was clearly the turning point, and once Mum had been won round my father didn't even try to put up much of a fight. Popeye stayed, and in the end was much loved by both parents.

Popeye went on to survive much travelling around Tasmania by both motorbike and car (he much preferred the car!) and later flew to Darwin, where he died just before his 18th birthday of kidney failure.

Postscript: Popeye was definitely smarter than Jack (and Betty) – being the names of my parents!

Naomi Oliver
Leanyer
Northern Territory

3

Smart animals to the rescue

Team work

I'm nearly 90 so I'm not quite sure how long ago this occurred but it would be between 50 and 70 years ago. I'll give you the nucleus of the story.

It happened in Victoria and was reported in the papers of that time. It was on a farm and the parents couldn't find their little son (he was somewhere between three and four years old). The father could hear their farm dogs barking down at the dam. Feeling horror stricken, he rushed down to the dam to find one dog in the water supporting the little boy. The other dog was on the bank barking to attract attention. Both dogs were wet and had obviously been taking it in turns when one got tired.

You can imagine how the parents felt. They swore the dogs would always have a loving home. This really happened. A lot of people don't deserve their pets.

Thelma Williams
Clontarf
Queensland

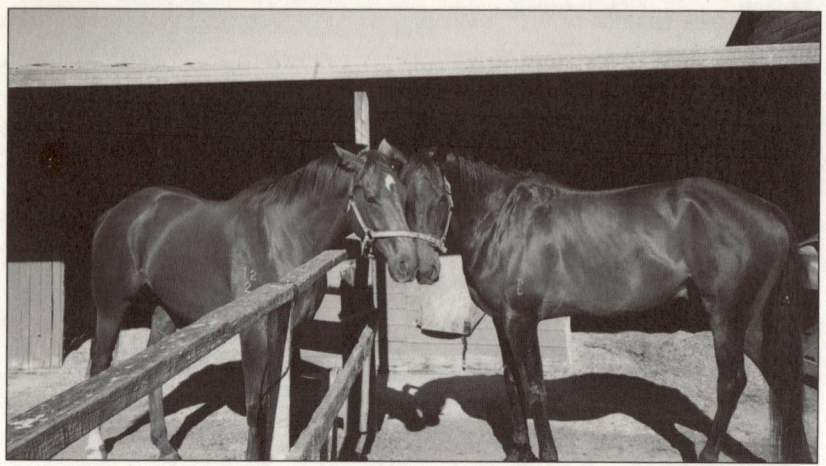

Sonny and a companion

He wasn't going mad!

I have a story about my then ten-year-old thoroughbred gelding called Sonny.

Back in February 2001, I had just stepped into our backyard to peg out some washing when Sonny began calling out and running round in circles and chesting his gate in his paddock (which faces the backyard). I sang to him, 'What's wrong, Sonny?' He continued circling behind his gate, then slamming into it with his chest.

I ran inside to get my husband, saying, 'Quick, there's something wrong with Sonny.' Sonny continued running round in circles at his gate until we got over to him.

Sonny then led us to the loose box in his paddock (which faces away from the house) and there we found Pav, Sonny's paddock mate, in a distressed state.

Sadly, Pav had been bitten by a snake and had to be put down a few days later after his system totally shut down. Throughout Pav's veterinary treatment Sonny stood by him with his head over Pav's neck.

It seems that Sonny knew Pav was very sick and was desperately trying to alert us by slamming into his gate when he saw me. We were amazed by his actions and think he is a very intelligent horse.

Brenda Watts
Catani
Victoria

Determination

Many years ago our pussycat named Fluffy showed amazing intelligence while we were out in the garden planting lawn.

I had put a pot on the gas stove to stew some fruit and had closed the back door as the baby was asleep in her bedroom. A short time later our cat began to pace up and down the back steps, mewing loudly.

For the first few minutes we took no notice, but soon realised he wanted our attention so we hurried inside and found the pot had boiled dry and smoke was filling the kitchen. Immediately turning the gas off and opening all the doors and windows etc, we took the pot outside, but the cat was not happy. He continued to meow at the baby's door until we let him in to see that the child was happily fast asleep and quite safe, then he went outside again.

We could not believe how wonderful that cat was and how determined he was to make sure our daughter was safe. The cat was quite famous around our neighbourhood for a long time.

Harold and Enid Massey
Singleton
Western Australia

Grace the dog

Guns are no joke

Grace is a 13-year-old boxer/heeler cross. She was a mere four or five when she showed this amazing behaviour to her owners, who were shocked and amazed at how clever, loyal and loving she was on that day – and still is.

One afternoon on a quiet weekend at home, Grace's owners Jesse and Kylie decided to sort out some boxes stored in the roof. Jesse was finding all sorts of interesting things from his childhood stashed in the boxes – like a time capsule. Grace grew tried of watching such boring antics and went to find somewhere comfortable to sleep. After a while, Kylie grew weary of looking in old boxes and retired to the lounge room to read with Grace.

Jesse entered the lounge room with great gusto, showing off a toy gun he had played with in younger years. There was much talking and waving his arms around as he retold stories of weekends and war battles won and lost. Unbeknown to Kylie and Jesse, Grace was watching this with great interest. Kylie was suitably impressed with Jesse's enthusiasm about his childhood stories and weekends of fun but disliked toy guns. She asked Jesse not to point the gun towards her as he told these stories. Jesse thought it was funny that a plastic gun – obviously not real – would upset Kylie.

As Jesse was in a jovial mood, he did just that and pointed the toy gun at Kylie for fun. At that point Grace took it upon herself to take action, and leapt across in front of Kylie and launched herself at Jesse. She took Jesse's arm in her mouth and dragged it to the floor, where she landed. She continued to pull at his arm until Kylie assured Grace that she was okay and to let go of Jesse's arm!!!

It happened within the blink of an eye; Kylie and Jesse didn't even know Grace was awake! She was smart enough to know that this was a potential weapon, and risked herself to save Kylie! She was, of course, praised and rewarded for such an act of bravery.

Kylie Ford
Sunshine Beach
Queensland

My little fur person

Tyson is a five-year-old Staffy and my little fur person. We believe that he understands a lot more than he lets on. There have been too many situations for it just to be coincidence. Here are two of those situations.

Three years ago while moving house we tried to explain to Tyson what was going on. We involved him in clearing out the backyard of all his toys and bones. We showed him how we put them in the trailer, then we drove him to the new house and took them out of the trailer.

Smarter than Joey

Later, while still packing at the old house, we went outside and found a bone by the trailer. We thought we must have missed picking it up and went back inside. Later we went outside again, this time there were a couple of bones and toys.

Tyson seemed to know that we were moving and was digging up his secret stash of bones and toys and putting them next to the trailer for us to take.

The second occasion was on a hot night when we left Gerald, our canary, outside. In the morning I went outside to find that roaming cats had pushed over Gerald's cage. The plastic bottom was smashed and there were seeds and bits of plastic everywhere. I had a quick look in the cage and couldn't see Gerald.

Distressed and crying, I looked around the backyard for feathers or other evidence of what had happened to him, thinking the cats must have taken him. When I couldn't find any evidence I went back to the cage to throw it in the bin. Then Tyson came outside and I told him of Gerald's fate. He started headbutting the cage and barking at it. I lowered my head to have a look and saw Gerald's little legs poking out from under the seed bowl. I moved the bowl and Gerald popped up from underneath. We covered the bottom of the cage and took Gerald inside before he flew away.

Gerald had blood on his head and red legs, and still has a hole in his feathers from the fall. I'm glad that Tyson could see a feathery breakfast underneath that seed bowl. If it hadn't been for Tyson, little Gerald would never have been rescued. In my eyes, he deserves a medal.

Anita Burdett
Athelstone
South Australia

Mitzi, our black and white moggie

It was 16 January 1992 and we were all tucked up in our beds. It was just after midnight when it all happened.

Where we live is a very busy corner for traffic. Although we are elevated we can still hear the madmen screeching their tyres as they speed around the corner, and smell the tyres as the drivers attempt massive burnouts as they try to take off on a hill start – well, basically it stinks.

ANYWAY.

Mum's cat Mitzi was what I call Mum's protector and this night we all owe Mitzi our lives. For under Mum and Dad's house is a flat that at the time was being rented out. At about midnight Mum woke to the smell of something burning and Mitzi going crazy with her meowing. Mum thought it was just a car doing a burnout so she didn't think any more of the burning smell. As for Mitzi, Mum just got fed up with her. Well, this wasn't what Mitzi wanted so she ran out of the house and meowed outside my window, so of course I was awoken by a cursing cat.

The next few hours go by in a blur but I can remember Mum screaming, 'There's a fire, there's a fire in the flat!' Next thing my Dad and I are pulling on whatever clothing we can find and Mum is on the phone to the fire brigade. All this time Mitzi was checking to see (I think) if we'd got the message.

It wasn't long before the three fire brigades, two ambulances and I don't know how many police were there, as we thought the girl was still in the flat – as it happened, she wasn't. The girl had decided she had had enough of life and tried to commit suicide by setting fire to her mattress and had leaned it against the only doorway out.

Mitzi was not happy until everyone was together again and was well rewarded for the job she had done.

I don't think I will ever forget that warning meow for as long as I shall live.

Unfortunately, Mitzi died in December of 1992 due to nose cancer. It was like losing a friend and a mate.

Barb Smith
West Moonah
Tasmania

How old Charlie got our goat

During WWII, I was a lad on an English farm which adjoined a small market town. On the farm we had a very smelly, very cantankerous and very randy goat named Charlie. Charlie was normally kept in the young sow pig's paddock. He didn't really mind this, but when I took the van to feed the pigs Charlie would go out the gate and straight into the driver's seat, sitting like a dog. I would then have to push him over to get in, and he'd then slaver all over my arms while I drove to feed the pigs. I'd have to heave him out on the way back.

Charlie found a way to escape the paddock and would often slink off into town. Once, he bailed up two old ladies in the church doorway and chewed his way through the flowers destined for the harvest festival. Luckily, the ladies got out and informed the local law. They, in turn, would ring us: 'Please, Mr – come and get your bl**dy stinking goat' from wherever he was.

Oh, old Charlie wasn't at all particular. He'd amble in any door which happened to be open – baker's, barber's, the pub, it didn't matter much. He'd slaver all over anyone who got near enough – his pong was so overbearing that it would hang in clothes for weeks.

My brother often had to pay out for cleaning costs – clothes were on coupons and hard to come by. I was the one who usually had to go and get the foul-smelling old miscreant; I'd grab him by the ear and march him back to the farm.

We tried fitting him with a collar and eight feet of chain with two 56-pound weights; surely this would stop his gallop, but no. He found a way of wrapping the chain around a post and slipping out of his collar and he'd be off again.

There was an old Irish lady who used to drive her pony and trap 16 miles just to have her nannies served by Charlie. She pleaded with my brother Dick to sell him, but Dick was once told by a Welshman that if there was a billy goat on the farm that farm would never get foot and mouth disease. The farm adjoining us got it in 1936.

Now, Charlie was really antisocial towards American servicemen in khaki. Perhaps no one had told him that they were supposed to be on our side. If he saw them in town he'd get downright stroppy, butting them (he had no horns), then turning round, pellets coming from his bum like a machine gun. It's funny, he never did anything to me when I was in my Army Cadet and later my Home Guard uniform.

One night, coming home from parade around 10 pm, I found old Charlie loose outside the big hay barn that was often used by the Yanks for, among other things, a kiss and a cuddle. He must have smelt the cigarette smoke, as he bolted around the other end of the barn and launched himself into the barn door – and at that precise moment the floor erupted into a wall of flame. There were shrieks and screaming. The guys came out pulling up their strides; girls and some married women, very sheepish, with knickers in hand. I knew most of them, coming out covered in burnt particles of hay.

Old Charlie had probably saved their lives. I grabbed him by the ear (none the worse, except for scorched feet) and put him in an empty shed. The Yanks helped to put out the fire with buckets and a stirrup pump.

I'd long left the farm when I was told that Charlie had finally succumed to a .22 magnum bullet after falling into a sewage trench and breaking his back. He was buried where he died.

Alas, the farm is no more. Where Charlie once held sway, houses stand, the urban sprawl has taken over. But one can rest assured that the beady-eyed old sod's ghost will be hovering around somewhere looking to make mischief.

Bill Daking
Kingsley
Western Australia

A clever mate

This event occurred in 1974 and this is exactly how it happened.

It involved two pet sheep, one a wether called Blue Tail (so called because, as a lamb, he had a Raddle mark on his tail). The other was a small ewe called Wobbly. She was born with a soft protrusion at the top of her head that resulted in slight brain damage, which affected her coordination.

When Wobbly walked she wobbled. Sometimes, when she wanted to go one way her body would go the other, and she'd occasionally roll over and become cast. Wobbly was happy though and reared beautiful lambs. Both sheep were a few years old and lived in a five acre paddock with other sheep.

One particular morning, Blue Tail – who never refused a feed of oats – came up to the fence by the shed where I was working and stood there baaing at me. I told him he was greedy and did not need any oats. He kept calling to me. So I gave in and put some oats over the fence for him. He ignored them and continued baaing at me. When he wasn't getting any reaction from me, he started walking away. He'd take a few steps, then stop and look back at me and baa again. So I got through the fence and began following him. He kept glancing back to make sure I was still there, at which time he would stop talking.

He led me to the other end of the paddock, where I saw Wobbly, cast, legs straight up in the air. Blue Tail walked to about ten feet from her and stood still, apparently to make sure I'd seen her. As soon as I walked over to help her, Blue Tail just ambled off the other way to eat grass.

And many people say sheep are dumb!

Sally Stanley
Digby
Victoria

The sheep's godmother

We call our horse the sheep's godmother. She protects them from the foxes that prowl around our paddock, stands over a dying sheep to keep it company, refuses to let animals or people near a newborn lamb and its mother for two days and generally takes care of the flock.

The best story happened one day in August. My daughter had been trying to jump the horse over a board on two bricks but the stubborn horse kept running round the side of the jump. We had a paddock cordoned off from the animals as it was under crop and the fence was about three feet high. We put a board across the gate to stop the horse getting in and eating too much but to allow the sheep under for a short time to feed. All the sheep had lambs, and the lambs were separated from their mothers in the tall grass.

Mothers and babies started to cry and run around looking for each other and there was terrible confusion in the paddock. On hearing the crying, our horse went to the bottom of the adjacent paddock. She ran uphill for about 50 metres, jumped the fence (which was five times higher than the jump my daughter was trying to get her to jump) and ran round the paddock from the outside, running in smaller and smaller circles so that, in the end, all the sheep and lambs were in the middle of the paddock and the lambs found their mothers!

I would not have believed it had I not seen it for myself.

Rhondda Cahalan
McLaren Vale
South Australia

Maximillion

Billy was a little boy who lived in a tin shack in the country with his Grandma and a few old chooks. Not a very exciting life for a young boy and, as you can imagine, Billy was very lonely. He spent his time looking after the chooks and his vegetable patch or going for walks in the bush.

Once a month his Grandma would get out the old horse and buggy and they would travel to the nearest town for supplies. They would leave as soon as the sun came up and wouldn't get back till the sun had set. Billy enjoyed the day talking to the townsfolk and, for a while, forgot that he was lonely.

On one such trip, while his Grandma was buying the groceries, Max, one of the locals, handed Billy a sack.

'Like a new pet? I found him wandering in the bush. Looks like he lost his mother,' he said. Billy looked in the sack. He saw what looked like a small-sized football with a rubbery, whiskered nose and small, black eyes blinking at him. 'What is it?' asked Billy. 'It's a baby wombat,' Max said. 'You'll really have to look after him because he's too young to look after himself.'

'Gee, thanks,' smiled Billy. 'My own little friend. I'll call him Max after you.'

Billy didn't quite know how to tell his Grandma about Max so he put the sack in the buggy. It was already dark as Billy helped his Grandma unload, and he was still trying to think of a way to tell her about his new friend when he heard a loud scream. He rushed into the kitchen and there was Grandma up on the kitchen chair with a befuddled Max looking up at her wondering what all the fuss was about. 'Get that overgrown rat out of here!' yelled Grandma.

'But Grandma, he's my new friend, Max, and I have to look after him. He's not a rat, he's a wombat.'

'Empty one of those grocery boxes and put that thing into it so I can get down. Tomorrow we'll take him down to the creek and let him go.' Sadly, Billy did as he was told. They sat down to a silent supper.

When Grandma went to bed Billy tiptoed into the kitchen to feed Max. He tried milk first, which he didn't like, so he gave him some bread and carrots. When Max had finished eating he lay on his back with his little, leathery legs sticking up in the air. As Billy tickled his tummy he drew in his little claws and was soon fast asleep.

'Billy! Get in here now!'

Billy woke with a start and ran into the kitchen. Grandma was back on the chair, and there was Max curled up in her rocking chair without a care in the world. 'Get the stupid wombat back in his box and lock him in the shed till I can get rid of him.'

'But Grandma, he's only a baby.'

'I've heard those things are covered in lice,' shuddered Grandma.

'No, he's clean,' protested Billy. 'He licks himself and combs his fur.'

'He'll grow into a nuisance and eat the vegetable patch and dig tunnels in the garden. Get rid of him! Lock him in the shed!'

So Billy went to visit his friend Max every day and fed him on bread, grass and carrots and Max got bigger and bigger, until he was so big that Billy knew it was time to return him to the bush. Billy put Max, who now looked like a fat little football, back in his box and took him down by the creek. He left him in the shelter of a log with a supply of food and sadly wished his friend goodbye.

That night Billy woke to the sound of his Grandma's screams. He rushed into her bedroom and turned on the light and there was Max pulling on his Grandma's blanket. Grandma jumped out of bed, yelling.

'Get that THING out of here – he nearly scared me to death!'

Billy picked up Max. 'You'd think he was a snake or a spider the way she carries on,' he mumbled to himself. As he was heading for the shed Billy saw a thin spiral of smoke in the bush, and through the trees he could see a red glow. 'Grandma, fire!' yelled Billy.

After the fire was out and all the excitement had died down, Grandma and Billy sat round the wooden stove drinking a well-earned cup of tea while Max squatted at their feet.

'You know, Grandma, if it wasn't for Max we would have been caught in the fire. He's a real hero. Don't you think we should keep him so he can look after us? I mean, just in case this happens again.' Grandma smiled. 'I think we can keep Maximillion.'

'His name is Max.'

'Yes,' said Grandma, 'but you have to admit he really is one smart Max in a million.'

Alexandra Kassos
Brisbane
Queensland

Beautiful Bella

My tale is as true as a tail that wags.

We have a beautiful golden retriever, Bella. We originally bought her from a local breeder known for breeding champions, so it was no surprise when Bella graduated top of the class when we finished her basic training in 1996.

In 1997 we introduced twin boys into our family. Apart from having to adjust to babies in bulk, we also had to quickly accept that we owned a dog whose level of awareness bordered on eerie.

At six months old our twins were at different levels of development. While Kyle was crawling and able to explore our home, Brodie was still only rolling around, playing with the toys I'd put down for him.

One day I was busy doing the washing, only checking on the boys intermittently as both were happy. From where I was in the laundry I suddenly heard a very high-pitched yelp. Bella has a very deep bark so I ignored it, assuming that it was the neighbour's dog. *Yelp.* I heard it again and it seemed very close. I immediately ran through to the kitchen, thinking that Bella had somehow caught herself somewhere or was hurt. What I saw unfolding was not what I expected.

Our back sliding door mustn't have latched properly. Kyle had managed to open it and was on his way out. Barring his way was Bella – crouched on her belly just outside the back door. Her eyes were rolling up, searching the glass, apparently for me. I stood quietly and watched out of curiosity. Each time Kyle tried to move forward Bella would lift her head from her front paws and yelp, just loud enough to startle him from moving. I was soon to discover that this episode was just the beginning.

Three months later, I'd taken the babies outside to play. Kyle was walking and Brodie crawling. I stood at the edge of our pergola and watched their activities, with Bella sitting companionably beside me. Kyle, of course, was happily teetering around our entire backyard, but

I began to become concerned when his attention turned to our large back shed. With a clearance of only about a metre, snakes are a constant concern for us in this area. I gave him the grace to wander away from it on his own, but he was soon heading behind it. I didn't say a thing, but thought to myself 'I'd better get him back'. Before I could even uncross my arms, Bella had left my side and was running down to the shed, only slowing to pick up a tennis ball in her mouth. I really couldn't understand what she was doing.

Bella quickly neared Kyle and, incredibly, gently pushed the ball into his tummy. He squealed with delight as he reached out for the ball, but Bella took a few steps backwards and placed the ball on the ground. Kyle turned around and reached down to grasp it, but Bella snatched it up and backed away from him again. My brain was truly trying to grasp what my eyes were seeing as I watched this performance for the next few minutes. Bella was slowly but surely enticing Kyle back towards our house. When she finally seemed satisfied that Kyle was out of harm's way, she allowed him to have the ball and calmly rejoined me.

This incident still amazes me. Bella's intention was not to play but to prevent Kyle from being harmed. Since that day, Bella has taken a very non-aggressive, protective attitude towards the boys. She often stands arrogantly between them and visiting dogs, not allowing the boys to even be sniffed by them.

As a family, we actually feel blessed to have our beautiful Bella and she is loved and appreciated for the 'work' she's done for us.

Madonna Wagland
Pittsworth
Queensland

Just like Lassie

Nelson, our beautiful beagle dog, was as devoted to us as we were to him.

One night, about two years ago, a sudden and powerful storm hit the eastern suburbs of Melbourne. Suddenly, the ceiling in our wardrobe began to balloon and fill with water. Wasting no time, my husband climbed up onto the steel deck roof at the front of our house to unclog the drain, which was causing the water to gush into the roof. He cleared the blockage and then stood up, silhouetted against the backdrop of the constant lightning, and called to me that he had cleared the drain. I could see he was feeling very proud of himself that he had saved all our clothes from a disaster.

Unfortunately, he wasn't able to enjoy his moment of glory for very long. The roof was dangerously wet and, as the rain cascaded onto it, the roof had become like a slippery dip. In one split second, he slipped off the roof and onto the brick path below.

I was terrified. My husband lay motionless in the pouring rain. I raced inside to phone for an ambulance. However, I must have screamed in fright, which alerted Nelson who, unbeknown to me, came running out of the front door, which I had left open when I raced inside. He ran up to my husband lying on the ground, nuzzled him, then dashed off in the pouring rain for help.

Our neighbours told us later that they had heard a noise above the sound of the storm and had gone to their front door to look out. Their front door was in an adjacent street to ours and they could not see into our front garden. While they stood there wondering if they should go out to look around in the terrible conditions, Nelson appeared in the light from their front porch. They told us later he was just like Lassie in the movies. *Please come and help!* he seemed to be saying, and waited impatiently for them to follow him.

Nelson and the neighbours arrived to help at the accident scene. Suddenly, the ambulance arrived and my husband was taken to hospital. Our neighbours comforted our shaken daughter and followed the ambulance, with her in their car. They stayed with us all until, miraculously, my husband was told that he had only sustained minor injuries. His lucky star had been shining that night!

The true star of that evening had been our beautiful dog. We were amazed at Nelson's ability to size up the situation and realise that his mum and dad were in trouble. He was a dog who was never allowed out on his own and always walked on a lead. His best doggy friend, a dear little Maltese shih-tzu cross, lived next door and he had been to visit on several occasions, but only as a treat for the two dogs.

How had he known to leave the warmth of his comfy couch and run out into the cold and the rain to our wonderful neighbours for help? He was always treated as a member of our family and he certainly never regarded himself as a dog, but I believe it took a very high level of intelligence for him to embark on his specific search for help on that night.

Our family was so fortunate to have a happy ending to this accident. My husband came home nursing a couple of broken ribs and a very sore ankle, but he gave his wonderful dog the biggest hug he had ever had in his life.

Suzanne Coomes
Canterbury
Victoria

4

Smart animals do strange things

Fair exchange?

While preparing for a garage sale, a neighbour dropped in with a carton of cuddly toys. Susie, my ten-year-old silky terrier, gave the toys a quick inspection. She decided on a white teddy bear almost as big as herself, took it by the arm, gently removed it from the box and placed it on her favourite mat. Susie then picked up her small, much worse for wear rag-dog, and placed it in the box.

> Daphne Maconachie
> Atherton
> Queensland

Spectators

I went up to Yulong Park at Medowie, which is not far from Williamstown Air Force Base, to watch my grandson play soccer. As I settled on the sideline I looked up to the far end of the ground and saw three large kangaroos hopping around. As the game progressed I noticed the kangaroos had stood erect and were looking at the game. They continued watching till the game had finished.

As I was folding my chair I looked across to the other side of the ground. The three kangaroos were hopping very fast down to the bottom of the ground. I thought they were heading towards some bush that was there. But I was wrong. There was a game of soccer going on between six- and seven-year-olds. I couldn't believe what I saw. The three kangaroos had stopped hopping and, standing erect side by side, were watching the game.

The locals didn't take any notice of the kangaroos, so I thought they must be regulars.

P.S. I can verify this by getting statements from the parents and the boys who played in the team on the day. At my local bowls club I told them about the actions of the kangaroos and one bright wag wanted to know if they were waving the Aussie flag!

Arthur Hiskens
Cardiff
New South Wales

A novel solution

One day I noticed a fox picking tufts of wool off the wire fence in a few different places. She ended up with a small ball of wool, which she carried in her mouth. The wool was from sheep that jumped through the wire of the fence.

The fox then trotted down to the Mortlock River and backed into the water. She gradually kept backing deeper into the water until all the water was above her mouth and nose and the ball of wool. Then she let the wool go and it floated away.

I grabbed the wool a little way down from where the fox had let it go and noticed that it was full of fleas.

Frank Watson
Singleton
Western Australia

A chook enters Ernie's dog kennel

Ernie's business deal

Cautiously, our two young pullets approach the dog kennel. Ernie, the dog, is inside. He shuffles over to make room and the hens tiptoe in.

Our free-ranging chickens have chosen the doghouse to lay their eggs in, rather than the nesting box that we provided.

Ernie – sporty, curly-tailed and button-eyed – seems happy to share. It is a cold morning after all and there's enough room for everyone.

My partner, though, suggests that they've negotiated a deal. It seems Ernie's cottoned on to the added benefits. It is a week or two before we can persuade the chooks to lay in the intended place. Meanwhile Ernie keeps his kennel doors wide open. Sometimes he would vacate his home for the hens' use, opting to bask on a bale of straw until they were done.

Without exception, if we didn't get to the eggs just as soon as they were laid, someone else did!

Heidi McElnea
Chewton
Victoria

Ding dong

We were looking after the Marcuses' cat at their home, while they served a four-year work commitment overseas.

Sinbad, or Sinny as he was known, was not a very friendly cat – a long-haired black Persian whose regular brushing and removal of knots required a sturdy pair of gardening gloves. His normal routine was to come inside at the end of the day, eat his meal, then head for the back door and return to the garden until bedtime.

One particular evening we had visitors. Our excited chatter took precedence over the feline feed time and, while we managed to prepare Sinny's meal which he ate in his usual manner, we just didn't hear his requests to open the door and let him out.

Our front doorbell had a set of chimes, with the various sized pipes running down the door on the inside. In exasperation, Sinny reached up and belted the pipes with all his might, the doorbell rang, I opened the front door and out he shot.

I was amazed that he had worked out not only that when the doorbell rang the front door would be opened, but also how to make it ring, as this was not his usual exit point.

Patricia Clarke
Belair
South Australia

Stupid dog!

When my Lowchen, Jake, was a puppy I trained him to retrieve balls and he has become a 'retriever' in all but name. When we go to the park all he desires is for me to throw his ball, so that he can bring it back and start all over again!

Recently, we were at our local park where a gentleman was throwing a ball for his dog. The gentleman and I started to walk around the perimeter of the park together chatting, both throwing balls to our respective dogs. Though the other dog would chase his ball, he would not bring it back to his owner.

Jake was clearly perplexed at the stupidity of this dog. On the next occasion when we both threw our balls, Jake raced away, selected 'their' ball, carried it back and dropped it at the feet of the gentleman. He then went back and retrieved his own ball, returning with a look of pure delight on his face!

Carolyn Brown
Killara, Sydney
New South Wales

Gilbert, my clever galah

Gilbert is my pet galah, though I often forget he's just a bird. Each morning, not long after dawn, he begins with *D'you want a cup of tea darling?* He is very persistent and asks over and over again with increasing volume – until I get up and make him one. He has a favourite cup and enjoys his 'cuppa' with a piece of wholegrain toast.

Recently, I discovered I'd left his cup outside the night before and, as it was a frosty morning, I offered him his tea in another cup. Gilbert was not impressed and refused to drink his tea until I finally relented and braved the cold outside to fetch his regular cup.

Smarter than Joey

Gilbert the galah and May

He gets very jealous if I spend too much time knitting or crocheting. The moment I put my needle or hook down, he grabs it and runs away to hide it under a chair – where I can't easily reach it.

If he thinks I'm spending too much time talking with a friend, he starts up such a screeching, squawking racket that it's difficult to carry on a conversation above the noise. My friends soon get his hints. Worse still is when he gets that neglected feeling when I'm on the phone. He's been known to chew through the cord on a number of occasions.

If someone knocks on the door Gilbert calls out *Come in darling*. His words have made countless people comment that they thought I already had a visitor.

Recently, he accidentally got out of his cage and flew off. I thought I'd lost him. My persistent neighbour went out looking for him. She called out to every galah she saw, hoping that he would come down so she could take him home to me. Gilbert must have had enough of freedom because he flew down and landed at her feet. He then proceeded to climb up her leg and onto her shoulder, where he remained till she got him home again. I heard him calling out *Are you there May?* And when I opened the door he greeted me with *What did you think of that darling?* Needless to say, I was so relieved to see him I quite forgot to scold him.

When Gilbert gets tired at night he says *D'you want to go to bed darling?* He keeps that up till I put his cover over his cage or turn out the lights.

Gilbert has a vast range of conversation, which he seems to be able to use in context. This makes him great company and he often has me laughing. He is much better than a dose of medicine.

May McFarlane
Victoria

Wolfgang travelling in the dinghy

Dinghy dunkings

When we started our cruising life from Fremantle, Western Australia on our catamaran yacht *Stray Cat*, we took our tabby house cat Wolfgang with us. He was ten years old at the time and had never been on a boat before, although he liked travelling in the car.

Our first stop was Shark Bay and to take Wolfgang in the dinghy for a walk on the deserted shore. We had a harness and long string in case he decided to explore the countryside and not come back. Although he enjoyed scratching on logs, digging in the sand and eating grass, he didn't like the harness at all. Wolfgang naturally wanted to stalk in the cover of the beach scrub, while we enjoyed walking on the exposed beach sand. This led to a battle of wills. Wolfgang would head to his side of a bush with us holding the lead on the ocean side. After a

Smart animals do strange things

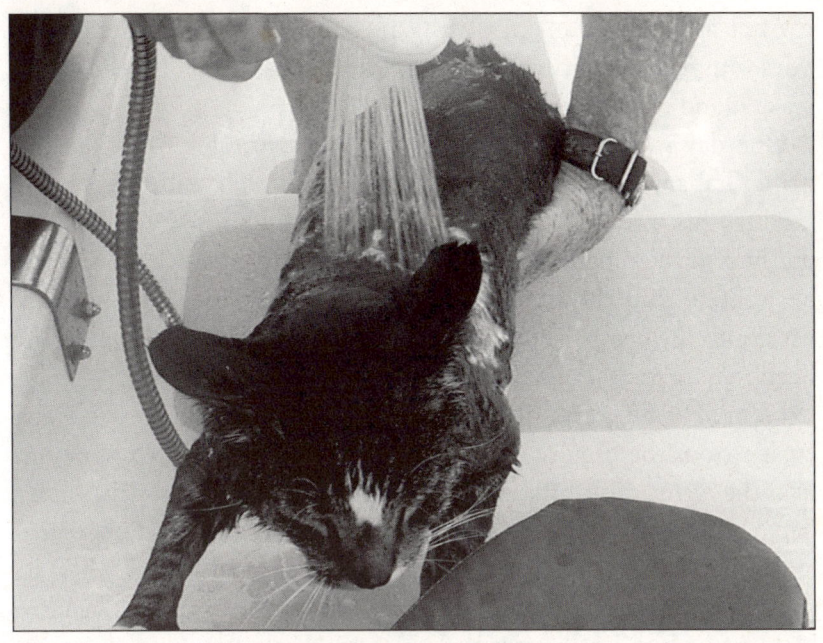

Wolfgang having his fresh water shower

big stand-off – you guessed it – we inevitably gave in and walked on Wolfgang's side of the bush. Once he established who was boss, he would grudgingly walk on our side of the obstacles for the rest of the walk. I think it was more like him taking us for a walk!

When returning to the yacht in the dinghy, Wolfgang was very eager to get on board. He developed an unrealistic perception of his ability to leap, and before we were close enough he would hurdle from the front of the dinghy, his front paws clawing the slippery deck, and fall into the water. He could swim very well and occasionally did a lap around the boat before we fished him back on board. This led to Wolfgang reluctantly having a freshwater shower to get all the salt off his fur.

After numerous dunkings Wolfgang began to associate the dinghy rides with the indignity of his unexpected swims. Sometimes he would make it and sometimes not.

One day while anchored in a bay inside Ningaloo Reef, my husband Graham took Wolfgang to shore for his walk and to get some sand for his tray. Wolfgang maintained his stubborn streak. Graham put him in the dinghy ready for the trip back to the yacht and, when he went to pull the dinghy down to the water, Wolfgang escaped back onto the beach. This happened several times. He had decided he wasn't going in that dinghy again!

From my vantage point on the yacht I was amazed to see Wolfgang escape from the dinghy once more, this time choosing to wade out into the water, with little waves lapping around his legs. This didn't deter him, and he started swimming out towards me on the yacht anchored in the bay. He had only swum about ten metres when he must have decided it would be too far for him to reach the yacht, so he turned around and swam back to shore. Wolfgang jumped straight back into the dinghy, very pleased to be taken back to the yacht in comfort. This dinghy didn't seem too bad after all!

We don't bother with the harness now, as we leave him on the beach to scratch, dig and relax while we go for a walk. When we return Wolfgang will always be waiting under the shade of a bush or rock near the dinghy, making sure he doesn't miss the boat. He now waits until the dinghy comes right up close to the yacht before jumping aboard, and most of the time he makes it! Wolfgang is 15 years old, very healthy from his fresh fish diet and loves cruising.

Meredith Sunderland
Gwelup
Western Australia

Smart animals do strange things

Trust

Sitting at the table one night I noticed a little tiny spider, about as big as the small fingernail on an adult hand, standing near my book.

He (or she) didn't look right, and when I got my magnifying glass out I could see that his front legs were totally bound together by a spider web. It was wound around tightly. It looked like he'd got his feet caught in a web and he'd spun around and around before escaping. Poor little bugger.

I got a needle and my magnifying glass and very slowly and carefully began pulling the web away from his front legs. His legs were much skinnier than cotton so, as you can imagine, I was very careful.

It took about half an hour to free him. At first he tried to escape, holding his little bound legs in the air, because he thought I was trying to hurt him. But after ten minutes or so he was hooking the web and levering himself back away from the needle and working the web off himself. The dear little creature was using the needle as a tool to free himself. Now, that is a smart little thing!

>Hannah Grace
>Mullumbimby
>New South Wales

Playing fetch

We once lived on a small acreage in Healesville, Victoria. We would walk our two dogs, Rocky the Rottweiler and Penny the collie/shepherd cross, around the property on a daily basis.

By chance, the elderly farmer living next door decided to put a baby calf named Charlie on his property to keep his grass down.

As we walked our dogs each day we would walk past the calf, and our dogs would run back and forth chasing sticks that we found in our paddock. To our amazement, the calf started to run up and down his paddock beside the adjoining fence, keeping pace with the dogs.

Then one day, Charlie the calf was waiting by the fence with a stick in his mouth as we came out to walk the dogs. At first we thought this was a coincidence. The calf then allowed my wife to take the stick from his mouth, she threw it, and he retrieved it just like our dogs, waiting anxiously for the next throw.

After that, the calf would be waiting on his side of the adjoining fence each day to join in with our dogs for his walk, and to play fetch the stick.

Garry and Jean Cowman
Croydon
Victoria

Togetherness

Years ago I had a black and white moggie, Felix. He was a birthday present on the Nullarbor Plains on my eighth birthday and grew up to be well and truly part of the family. Felix, however, took house-training to the extreme.

Whenever we went to the toilet, Felix would sneak in prior to the door being closed, and while we were sitting on the 'throne', he'd jump onto the bench, perch himself astride the hand basin and simultaneously pee down the drain. Needless to say, when visitors came we kept him locked outside.

Aren't animals funny?

Julie Gower
Brisbane
Queensland

5

Smart animals cope with life and death

A memorial?

This happened a couple of years ago. It moved me very much and I feel privileged to have witnessed it. My friend witnessed it as well.

While working in my paddock, I noticed that in an adjoining paddock, some distance away, there was a dead cow. About 15–20 other cows were forming a tight circle around the dead cow.

I stopped what I was doing and watched in amazement. The cows were keeping the young calves, who were obviously curious, from entering the circle by closing ranks and mooing at them to stay away.

Once the circle was tightly formed, with all the cattle facing inwards towards the dead one, the one and only bull moved to the dead cow. He looked, smelt and circled round it before resuming his position in the circle. One by one, the cows then moved forward and did the same, each resuming their position in the circle. This continued until all had done it. All the while they kept the calves away. This took about 50–60 minutes. They all then walked slowly away, at which time the calves had a quick look and a sniff and went away also.

It was as though the cattle were having their own memorial service and saying their goodbyes. It was truly amazing.

Gai Hovey
Trentham
Victoria

Patches and Commander

A tribute to humour and courage.

Where do I begin? I have been blessed in life, having been given the privilege of sharing 20 years with two beautiful horses. I'll start with Patches as his personality and intelligence amazed me constantly, let alone his incredible humour.

The day I got him I was 14. At the time I had no idea that he was already 20 (in human years). There were many people watching the grand ride to the bottom of the paddock and back – to see if we were compatible. We made it to the bottom of the paddock and, as I turned Patches to make our way back, he did an unexpected buck. It wasn't a huge buck or a meaningful one, but it was enough to dislodge a surprised rider. Where did I end up? Well, at the bottom of that paddock was a manure heap (and not a small one) – need I say more!

Patches, instead of running away as most horses would have done, stood there, looked at me, neighed – almost like a snicker, possibly laughing – then sighed. So I picked myself up and yelled to the worried onlookers, 'Yeah, I'll have him.' I could never have imagined the incredible life we would share from that moment on.

Many years before I owned Patches, he had a bad experience with kangaroos. A group of kangaroos went through the paddock he was in and ended up cornering him in the fence – he never forgot.

One morning we went out on a bush ride and, lo and behold, there they were – kangaroos. At the time, we were leisurely trotting up a track. Of course, Patches stopped fast – and me, well, I just kept going, straight over his head, and landed on my back. It was the worst winding I had ever had. Of course, I was unable to breathe or speak as I was trying to catch my breath. I could hear Patches running around the bush trying to find me, neighing frantically. This went on for about five minutes until eventually he made for the neighbour next to where he was stabled.

Smart animals cope with life and death

Commander and Patches – two peas in a pod

It struck me as funny that he never went home. Although I knew that nobody was around at the house during the day, did he know that? When the neighbour, a lady, finally came out of her house she saw a distressed horse pawing at the ground. He started to walk back across the road as if to say *Follow me*.

The lady caught up to him and grabbed the reins, just as I came waltzing out of the bush. I called his name and he neighed, so I guess we were both satisfied that we were okay.

The great thing about Patches was that whenever you fell off him he never ran away, but rather he would stand over you till he knew you were okay. He was older than me in human years, after all. He was very much like a parent – he certainly acted like one. And he did the same when Commander came along. Commander was such a softie that Patches would step in and fight all his battles for him in the paddock. The other horses seemed to respect Patches' age also.

Patches was so unique that people were astounded when they either witnessed or heard about the quirky games he liked to play. His favourite was wrestling. You grabbed him around the neck and he would push you down so that he had you as low to the ground as possible – so that you couldn't get up. Then he would bring his front leg forward, wrap it around yours, then pull his leg back so that you ended up sitting on the ground laughing your head off. Bear in mind that this was something he started. It wasn't taught to him, he did it of his own choosing. If he didn't want to play he wouldn't.

One day at a show, Patches was eating his lunch while I was showing Commander. He was tied up to the float. It seems that once lunch was finished Patches wasn't going to hang around for me to come to him, so he decided to come to me – he was good at untying safety knots. I can still remember the man reading out the winners of the previous event over the PA system, when mid-sentence I heard him say, 'Oh my god, where did you come from?' Yep, my Patches had stuck his head into the announcer's window and neighed at him. He had nearly fallen off his chair.

One morning I slept in until after 10 am. The house I was living in had no fence around it. The horses could come right up to the windows. I knew Patches could let himself out of his night yard if he really wanted to, but he never did normally. As I had slept in, and as he wanted breakfast, I guessed he thought enough was enough. I awoke to a tapping on the window. Thinking there was a person there, I got up and looked out the kitchen window. There were Patches and Commander, neighing at me – somewhat disgusted, I suspect, with the delay in breakfast. The weird thing was, I didn't leave halters on my horses at night-time, so what was the tapping on the window?

Patches' big front teeth, I suspect. There were slobbery marks all over the window. I guess horses will try and communicate with us by whatever means they can, and in a way that we can understand.

You know, I have so many stories but I cannot possibly tell them all.

But I need to bring Commander into it now. We lost Patches in 2000 with a tumour, aged 40 (human) years. In his last year, Commander fought all his battles in the paddock for him. It was really a sweet gesture and a tribute to their special friendship.

When they came to bury Patches, I grabbed Commander and we both walked to where our favourite riding places were. I didn't want Commander to see what was going on, for there was only one other who felt Patches' loss as I did and that was Commander. He was given some herbs for his grieving as he became very introverted. I remember, as we walked to all of Patches' favourite spots, how I stopped and asked Commander what we were going to do without him. Commander just stared straight into my eyes the whole time. We must have stood there for a full five minutes, looking at each other. Then, as the tears started to fall, this special horse of mine – who had really just lived in the shadow of Patches because of his strong personality – tucked his head under my arm and didn't move until I did. We were one in our grief.

Commander was never the same again. He was much quieter, but his loving nature was so strong. He was so soft and gentle, everyone loved him. If I sat out in the paddock with him he would play with my hair and ever so gently nip my shoulder – a typical grooming practice of horses.

Unfortunately, Commander's health seemed to suffer after that terrible day when Patches passed away. I believe that he didn't want to be in a world without Patches and was really only hanging around for my sake.

The last 14 months were the hardest as he fought with a paddock injury that left him partly paralysed. So many people were inspired by his courage, even when his treatment was a painful process. Commander knew all along that everyone was on his side and was trying to maintain his quality of life. He would even walk past his night-time feed, go outside to his yard and immediately lean up against the fence that we used for his therapy sessions – and there he would wait for his massage.

To everyone's surprise there was never a bite or a kick. I guess we all found out that he had a personality just as big as Patches', only in a very different way. Commander passed away in November 2002. Although it's been devastating for me, that was one horse reunion I wish I could have seen.

The horse world still has much to learn. We seem to be quick to berate instead of listening to the snippets of personality that horses are trying to show us. Many see the smallest of things horses do as bad manners, when indeed that may not be the case. Remember, work with, not against. More importantly, NEVER THINK YOU HAVE THE RIGHT, ALWAYS ASK FOR THE PRIVILEGE.

Karen Mol
Balga
Western Australia

Devotion

I am writing to tell you about my Tokenise female cat – eight years old at the time of this story (last year from 16 September to 29 October). Her name is Houdini (Houdi for short) – the second – and she isn't a real house cat. She sometimes will only come round for her meals.

My husband was diagnosed with terminal cancer and deteriorated quite quickly (he only lived a further six weeks). From that time on, Houdini hardly left his side. My friends and I marvelled that she only ever left the house to answer the call of nature.

She was either with him in the bed or by his side if he was sitting up. When the Silver Chain nurses called, in the last two weeks of my husband's life, Houdini would look at them – as if to say *Don't you hurt my master*. Everyone was amazed at her, we are quite sure she knew her master was dying.

My husband died at home, and Houdini went missing. As you can imagine, I was devastated. I had lost my husband, and now my precious cat was gone. She was away for about three days before she came home, quite bedraggled.

My friends and I thought that her disappearance must have been her way of coping with his loss. For some time afterwards she seemed to be pining for him. I have had a Burmese cat prior to having this Tokenise, and I find they are very much a one-person or two-person cat and very faithful and loving. She is back to her old self now and comes and goes as she pleases, but I often think of her devotion to her master.

Edna Kusen
Margaret River
Western Australia

Saying goodbye

Kozzy was our beautiful pet cat. She was white with grey patches. Her favourite spot was at the back of the vegetable garden, where she would curl up on an old wooden box and soak up the sun. At night, she slept snugly in our lock-up garage in her bed on top of an old armchair. She didn't like to be constantly patted and picked up, but on cold evenings when I was watching TV she loved to curl up on the sofa next to my feet.

Kozzy had to have three major operations in her life. The first was after she was desexed. The wound became infected and she had to be operated on again to remove all the infected tissue. This time it went well and she recovered. The next one was on her knee. My dad found her in the garage one morning unable to walk and in pain. The vet informed us she had a broken knee. He couldn't say how it happened but he was able to do a reconstruction to fix it.

Then, when she was older, we found out she had cancer on the tip of her nose. We had been trying to keep her out of the sun as much as possible once we learned that cats with pink noses were susceptible to cancer, but the damage had probably already been done. The vet recommended having the growth removed. I was so upset. I knew that Kozzy hated going to the vet and now she had to have another operation.

When I picked her up after the operation, the nurse asked me to try getting Kozzy out of her enclosure as she would not let either her or the vet pick her up. As we walked towards her enclosure I heard her hiss, but as soon as I called out to her she quietened down and let me pick her up. She knew she was going home. I hoped it was true that cats had nine lives because now Kozzy had used up three.

A couple of years passed, then when she was 12 years old she began to limp slightly. I thought it may have been a strain and that it would heal, but after a few weeks it was just getting worse. She had to go to the dreaded vet again. (Our vet was very nice, but Kozzy hated going in the car and maybe she was like my dad who hates going to see doctors.) As soon as the vet said he suspected a tumour in her leg, I burst into tears. An X-ray confirmed the diagnosis. It was a bone cancer in her back left leg.

The vet gave us two options – amputate the whole leg or have her euthanased. I didn't want to put her through another major operation at her age, especially when the vet said it was likely that the cancer had already spread. A second opinion from the RSPCA clinic was the same.

Our vet had recommended having her euthanased fairly soon, but it didn't seem right to end her life when she was still enjoying her days. She still purred when I patted her. She ate well and went for her walks around the backyard. We decided to wait for a while and see how she went.

I began bringing her into the laundry to sleep at night to make sure she stayed warm, especially since it was the start of winter. Towards the end of winter, we noticed that she didn't walk around as much. On nice sunny days I would put her bed outside under the patio, where she could enjoy being outdoors.

Then, in early September, Kozzy's health suddenly deteriorated. She had no energy. She was losing her appetite and only getting out of her bed if she had to. Within a few days she was hardly eating anything. I knew the time had come. I made an appointment to take her to the vet clinic on the Saturday morning.

On that morning, I hoped that she would stay quietly in her bed, but I could hear her meowing to be let out of the laundry. I wanted to get her into the car without a struggle. (We had a foam box to use as a lid over her bed so I just needed her to stay in her bed.) She was already up, so I opened the door. She went straight to the back door and waited for me to open it. I let her go outside. Maybe she was starting to feel better, because normally she would wait for me to carry her outside. I thought of cancelling the appointment.

It was a sunny spring morning and I watched as she walked towards the garage. She slowly walked through the garage, making little stops to have a sniff. She sniffed the car tyres and the legs of an old table. She went out the small door, to the back of the vegetable garden and looked around for a couple of minutes. Then she began making her way back through the garden, sniffing all the plants as she passed them.

As my parents and I watched, I suddenly realised that she was saying goodbye to her home. It was her last chance to walk around her backyard and it was as if she knew it. Maybe she was also letting us know that the time was right for her. I placed her bed just outside the back door and she climbed back in and curled up into a ball. We let her rest for a while, then Mum and I took her to the clinic and stayed with her till her last moment. It was the hardest thing I've ever had to do. We brought Kozzy home, found a nice box to put her in and buried her in her favourite spot at the back of the vegetable garden.

Maria Alberico
Mulgrave
Victoria

Changi introduces the next generation

Changi was a desexed male cat. He had been with my family for over 13 years. He was totally black with no other markings of any kind. He could hardly be described as an 'inside' cat; he spent most of his life roaming outdoors. Like many other cats he would often meow at the front door until someone, usually me, investigated the problem; at which time he would present a half-stunned, but still alive, mouse. With all his faults he was a very faithful family cat. I might add that we had relocated twice since he arrived on the scene as a kitten.

One day, when Changi was about 13 years of age, I heard him doing his usual meow at the front door, and of course I investigated. This time it was different. Instead of a mouse, there on the planter-box alongside the front door was a very small kitten, jet black in colour with no apparent markings. It marched up and down the planter-box with its tail held high. Changi was meowing away and drawing my attention to his new-found friend. Well, we didn't need any additional cats to add to our household pets and I admit that I shooed the kitten away.

Twenty-four hours later, our next-door neighbour called to me over the fence and told me that she had found Changi dead under a hedge in her garden. At 13 years he had died of natural causes and was given a very decent resting place in our rose garden near his old friend, our boxer dog Leah, who had died the previous year aged 15 years.

I now pose a question: did Changi bring along his replacement, knowing that his time had come? But how does a cat know to bring along another one of the same colour and maybe the same gender? Why didn't he just present a brown one, or a tabby or whatever? Is it really possible that cats have the instinct, or sufficient intelligence, to know that they are about to die and to have the decision-making ability to seek out a suitable replacement? I don't know!

It pleases me to think that Changi had had such a good life with our family that he didn't want us to be 'catless' for too long. I really wish we had kept the kitten, but we didn't. We are not that clever, or as intuitive as our feline friends.

I know that cats have been venerated and given special status over many years. Ancient history provides many tales of cat-ability. So why am I surprised? Do you have an answer?

Alex Copeland
Scarness
Queensland

Smarter than Joey

A hatchling Rosenberg's goanna

Clever solution

The female Rosenberg's goanna found on Kangaroo Island has to find a good place to lay her eggs. The eggs need to be kept warm.

For thousands of years, this goanna has used termite mounds to do this job for her – but not just any termite mound. She goes to a few termite mounds at night and scratches around near the base. Then she comes back the next day to see which mounds have been repaired. This means there are still termites in the mound.

When she is ready to lay her eggs, the goanna digs a special hole just big enough to crawl through and then lays her eggs inside the mound. The termites close over the hole in a few places for safety. The mound stays warm enough for the eggs, even at night.

After the baby goannas have hatched, they stay inside the mound for a few weeks and then crawl out through the hole made by their mother.

(Acknowledgement: Mike McKelvey at Pelican Lagoon Research Station, Kangaroo Island.)

Stephanie Mallen
Belair
South Australia

The last farewell

I was living in South Australia a few years ago at Christie's Beach when this happened.

My elderly neighbour and I were sitting in the front seat of our car on the esplanade having some lunch, and naturally the gulls were hanging around for any leftovers. Suddenly, a maniac of a motorist came along, scattering the gulls and hitting one, which fell to the ground dead.

What happened next was an awesome sight. In a matter of minutes the cries of the distressed gulls brought others from the south (Port Norlunga and West O'Sullivan Beach) – there were dozens of them. They circled the body (around and under power lines) three times and then dispersed gradually, until just one gull was left. It then circled the dead gull and left, leaving hardly a gull in sight.

I have told many people of this, but have never put pen to paper until now. So what do you think? Was it a farewell to a mate or a last salute?

I am 83 years old, but I can honestly say I have never seen anything that has stuck in my memory so much.

Audrey Broadstook
Wentworth
New South Wales

A great loss

Quite a few years ago we had a very special 13-year-old dog (Pomeranian/Aussie terrier cross) named Sheha – she was such a loving, loyal family dog.

One day my mother bought Sheha a plastic toy squirrel that 'squeaked' as you pressed it. She seemed to love it right from the beginning – the toy was as big as your hand and coloured white. From the very first day she would carry the toy squirrel everywhere, holding it by the scruff of the neck and being ever so protective of it. The toy would squeak as she picked it up – she loved it *so* much!

A few months later the much-loved toy lost its squeak and did not make any noise at all. Much to our surprise, Sheha was saddened by this – she was obviously so in tune with the sudden silence that it wasn't too long after that she buried the squirrel toy in the back garden.

This was such an unbelievable act – we always said she cared for that toy as if it were her very own pup. She took good care of it, not leaving its side, up until the very end when she knew what had to be done. Her baby had passed and she farewelled it goodbye.

Another few months later Sheha fell pregnant for the first time and gave birth to a few pups – all had died except one. I was happy to see that Sheha had become a dedicated mother again – and very much happy at that too!

Madeline Stathopoulos
Avonsleigh
Victoria

6

Smart animals think of their tummies

Watch and learn

I never really thought too much about possums and whether they were intelligent or not. But one night one particular brush-tailed possum certainly made me sit up and take notice of the species.

I was camping out in Croajingalong National Park at Tamboon Inlet and we had transported our food via kayaks in a large watertight barrel with a screw-top lid. In the middle of dinner this very bold possum lingered around us, obviously after some food. It approached the barrel, realising this was where the treasure lay, and tried to bite and scratch the barrel to get to the food. We then interrupted the possum to fetch some food out of the barrel, screwed the lid back on and sat back down.

To our amazement the possum reapproached the barrel and, instead of scratching and biting the barrel, it now tried to grab the lid with both paws and open it just as it had seen the humans do! It didn't have any luck as those little paws just couldn't do quite the job that was necessary, but it certainly gave it a good try.

Jenny Hourigan
Mordialloc
Victoria

Gypsy the horse

Sweet Gypsy

Gypsy was 14 years old and she was bombproof. I had her for about 12 months to two years before I knew she was smarter than other ponies.

We used to tether her during the day in a big paddock and lock her up at night. Well, this particular morning I went to let her out and she was already in the paddock. She had opened the gate locking her in, then shut it behind her and walked through the gateway and into her paddock. This was in a quiet street in Warragul.

Gypsy would get a cup of coffee, hold it between her teeth and lips, and then tip her head up and drink the contents. The cup had to have coffee, white, with two sugars in it before she would drink it. She would know the difference. If you didn't put sugar in it, she would stick her tongue in the cup and do a taste test. If she didn't like it or it had no sugar, she would push the cup over with her nose.

We had an office down the yard where she was locked in at night. We would leave the door closed when we weren't there. One morning Gypsy decided to open this door with her mouth, walk in and look around for the sugar (which was on the bench in a Moccana coffee jar). Gypsy saw this jar, gently pushed it over, and licked, rolled and pushed the lid. About five to ten minutes later she had the lid off, then she gently pushed the sugar out of the jar. Then, when she got to the bottom, she pushed it over again.

My pony was the best; she thought she was a dog. She would follow you around without a head collar or lead rope, and then if you stopped she would headbutt you in the back or bottom to get you walking again. If I ever bought my lunch when we went riding, you would have to buy Gypsy a meat pie with sauce and a bottle of coke. If you didn't, she would try and eat your lunch and drink your drink.

I just think Gypsy was the best horse ever.

Rachael Osborne
Cranbourne
Victoria

A cunning breed

My stories relate to an Appaloosa horse. This breed is renowned for being smart, but even these incidents amazed me.

I actually have two horses, an Appaloosa named Billy Jack and a pony named Mickey Mouse. At various times I have had to divide their paddock with electric tape to stop them eating too much grass and getting too fat. I only used a portable electric fence unit with orange tape, and sometimes this was left in place but not actually switched on.

The mystery started when I noticed that often in the mornings both horses would be on the wrong side of the tape eating the long grass. I could not figure out how they got there as the tape was still up and seemingly untouched. I thought maybe Billy Jack had jumped over it, but I knew Mickey would not have done this as he hates electric fencing and is also too old to have jumped it.

I decided to set it up one day and hide to watch them to see how they did this. To my amazement, Billy went up to the tape and tested with his whiskers whether it was on or not. If it was on, he turned away and stayed where he was supposed to be. If it was off, he put his back hoof on the tape to hold it down on the ground. Mickey would then walk across it to the fresh grass, then Billy (keeping the hoof holding down the tape in place) sidled around until he also was on the other side. He then walked off and they both had fresh grass to eat.

The second story is about Billy opening farm gates. When we first moved to this property I was awakened early one morning by the sound of someone heavy walking along the decking around the house. I went outside, to see Billy walking along it looking in the windows. I thought someone must have left the paddock gate open, until later on I saw Billy actually opening the gate by lifting off the chain with his lips. Needless to say, we changed all the gate fittings to ones he couldn't do this with.

All except the front gates to the property. These had the sort of catch that just flops over from one gate onto the other one. I had come back from shopping one day, closed the front gates and let the horses out to eat the grass around the house. I went back to the car to start unloading the shopping and heard something at the front gate.

I turned around, to see Billy standing at the front gate but looking back at me to see if I was watching. When he saw that I was, he just stood there looking over the gate. I thought things were safe so started taking the shopping inside.

I had just got in the front door when I heard the gate latch clink, so I looked back to see Billy and Mickey taking off through the gate down the road. He had actually waited till I had gone inside before he opened it! Luckily, they only went for a short run and then stopped to eat the grass at a neighbour's place. I jumped in the car to go after them and when I caught them up, Billy walked over to me as if to say he had had his fun and was now ready to come back home. I led both horses back, with their lead ropes through the car window and them trotting alongside the car. So now when I have them around the house, I have to tie the gate latch down so he can't flick it open.

So, a warning to anyone who owns Appaloosas: beware, because sometimes they can be too smart for their own good.

Koo Wee Rup!

Pam Cartledge

In a safe place

We have a sunroom overlooking our back garden. In the garden we have a large bird feeder hanging from the branch of a tree. A great variety of birds use this feeder, including parrots, twenty-eights, galahs, bronze wing pigeons, magpies, butcher-birds and even crows.

One day I bought a punnet of petunias and planted them out in our backyard flower bed. I then put some chips left over from lunch on the bird feeder. The first bird to arrive was a crow. It gorged itself on the potato chips. When it couldn't eat any more, it took a chip in its beak, flew to our flower bed, put the chip on the ground, pulled out one of the petunias I had just planted and stuck the chip in the hole. The crow then flew back to the feeder and repeated the process twice more. Each time, it pulled out another petunia seedling. The crow then flew away.

About ten minutes later a crow came back, pulled the chips out of the ground and ate them. No sir, I'm not going to tell you that it replanted my petunias. Of course, I had no way of knowing if it was the same crow, or if the first one had told its mate where to find a feed.

Bert van Leeuwen
Mandurah
Western Australia

My next meal

My wife Dot and I had been taken over by an Indonesian kitten, which adopted us in Tandjong Priok, the port for Jakarta. We were visiting that port in our nine-metre yacht, *D'Vara*, from Fremantle, Australia.

Tiger (he looked like a miniature one) soon settled down to life aboard the yacht and quickly adapted to the confined space. After visiting Malaya, Sri Lanka, and the Maldives and Seychelles Islands, *D'Vara* was heading north for Aden. Tiger had quickly learned that a patter on deck at night (he was normally confined below) meant a flying fish had landed somewhere on deck.

Every morning, when the screen covering the hatchway was removed, Tiger would race on deck and hunt for his breakfast. If he found a fish, he would carry it back to the cockpit in his mouth and drop it there before carrying on with his search. Sometimes he would

Smart animals think of their tummies

Tiger on the deck of *D'Vara* in the Indian Ocean

be lucky and find four or five fish, and all would be laid safely in the cockpit. After Dot had cut off their large fins, which he would not eat, Tiger would tuck in with every appearance of having been a successful hunter.

However, on the few nights when no fish had landed on board he would make several circuits of the deck, and even squeeze under the dinghy, which was lashed to the cabin top, to have a look there. When he finally realised there were no fish to be found he would jump into the cockpit, look whoever was on watch in the eye, and give a series of plaintive meows. It was quite plain to us that he was saying, 'Well, what have you done with my breakfast? I am sure I heard one pattering on deck last night. You must have let it jump overboard again.'

When approaching Cape Guardafui (now renamed Raas Caseyr) at the north-eastern corner of Africa, we sailed through what must have been an immense shoal of flying fish. Over a hundred landed on deck. When Tiger had carried as many as he knew he could eat at one sitting to the cockpit, he did not give up collecting them. Knowing they would quickly go bad in the tropical heat, Tiger started carrying fish down into the cabin.

These he neatly stacked in front of the refrigerator (a small kerosene-operated one), then turned to us with another meow, which plainly meant *Okay. Put them in there for later.*

His order was followed.

Previously in Malaya, we were away for almost a week, and the watchman we had employed to feed Tiger put down a great pile of fish on the first day, then didn't return. Tiger ate what was left there for him and nearly died of food poisoning.

S E Bradfield
Tully
Queensland

Prize catch

It was about 20 years ago when I found myself at one of the animal shelters crying my heart out for leaving a bull terrier there. I still remember that last look the dog gave me, and I knew its future was bleak as it was very much out of control. In my uncontrollable emotional state I decided to look around the shelter and visit all the animals awaiting adoption.

I took home a cat who was about seven months old. Not because I wanted a cat, but it gave me that sorrowful look and I thought I was obligated to put something back.

Muffins came home as a shy, timid cat with a nervous twitch and took a long time to settle in and take to us. When he finally settled in, his whole character changed and he developed a strong penchant for food that wasn't his!

Muffins stole food from the poor cat next door, even though he was well fed. These poor people could not speak English and had never owned a cat before, and he took full advantage of it. He stole food from the dogs next door too. They were chained and he managed to move their bowls just far enough out of their reach.

One night we were sitting in our lounge room and we heard knocking at the back door.

This was followed by a *thump, thump* and the cat's cry. We dashed out, and to our amazement there was Muffins trying to pull a whole frozen chicken through his little door and it would not fit. His prize catch!

Angela Thonbury
Dandenong
Victoria

Smarter than Joey

A bandicoot taking a hand-out

Besieged by bandicoots

I don't know if there is an animal equivalent of the old swagman's marks on the fence that alerted the gentlemen of the road to places that were good for a meal and a few bob. If there is, our fence must be covered with messages to the effect that 'Soft touches live here'.

It started with possums, then magpies, then bobtails and finally bandicoots. It progressed from an occasional crust of bread to a regular supply of peanuts. At times, as many as four bandicoots show up of an evening, all in search of a handout. They come in through the porch cat door and, if the dish is empty, they bang it on the bricks until someone comes out with peanuts. Two of the bolder ones will enter the kitchen and come into the living room to peer at us until we get up and find the peanuts.

The cats don't seem to mind the bandicoots, and on occasion I've found one or both cats eating Kibbles from one side of the dish while a bandicoot eats from the other.

Until recently, I wasn't aware the bandicoots had learned anything from the cats other than how to use the cat door and where to find the Kibble dish. The other night showed how much more they'd learned.

I heard something scratching on the front door, which pulled my attention from a not very interesting television show. Getting up to let Harley or Topaz in, I realised that both cats were already asleep on the sofa. What was trying to get in the front door? Switching on the outside light, I saw a small grey-brown animal crouched on the ground. I opened the door slowly. 'What do you want?' I asked.

With a small *whuff* noise, the bandicoot trotted into the living room. Its nails clicking on the tiles, it pattered across the room, through the dining room, and stood waiting by the back door. This manoeuvre had saved the beast from having to walk all the way around the house to its usual entry point.

'Great, now I'm a doorman for bandicoots,' I said, opening the back door and watching the little animal trot out onto the porch. There, it grabbed the edge of the empty dish and banged it on the brick floor.

I sighed and filled the dish with peanuts. The bandicoot began eating, looking over its shoulder at me every so often. I found a few wrinkled grapes and added them to the dish.

I returned to the sofa. One of the cats opened an eye, winked and went back to sleep. I sat pondering the saying 'Dogs have masters, cats have staff'. Obviously, this applies to bandicoots also. I am comforted by the thought that if St Francis ever gets door duty at the Pearly Gates, I'll be a shoo-in.

Karen Treanor
Mundaring
Western Australia

Oscar

Oscar was a golden retriever who, at three months of age, went to live with his new family in Tasmania. He soon fitted in with their daily routine and became a valuable and much loved family member.

His daily walks often took him to the local shopping centre and he would wait patiently by a shop doorway while his mistress completed her purchases. It did not take long for him to be warmly welcomed by shoppers and shopkeepers alike, as he greeted everyone with a wag of his tail.

His favourite shop was the butcher's, where his mistress always purchased a meaty bone. This was wrapped and duly presented to Oscar, who was only too happy to carry it home to be safely enjoyed in his favourite spot in the garden.

As Oscar grew older and wiser he was often allowed to take the walk to the butcher's shop alone to collect his bone. He never once broke the rules by stopping on his way back to enjoy a munch. The time came when the friendly butcher wished to retire, and the business was transferred to the newcomer. The old butcher agreed to stay on for a time to make the changeover easier and to make the necessary introductions to the customers.

One day when the door of the shop received a thump of a customer wishing to enter, it was opened by the new man. On seeing a lonely dog outside he hurried back to tell the old butcher. 'That'll be Oscar,' he told the new man, 'his bone is on the bench.' To the door again went the new man with the bone, only to return and say in a surprised voice, 'He doesn't want it, he won't take it.' Whereupon the old fellow turned to look at the bone and, in a scornful voice, said, 'Of course he won't take it, you did not wrap it.' So, neatly wrapped in clean white paper, the bone parcel was again offered to Oscar. He accepted it with a broad grin and happily trotted home, where his loving mistress was waiting for his return.

This story has been told many times to illustrate the fact that, besides being man's best friend, a dog is also a creature of habit and, trained in the correct manner from his early years, will always be willing to please.

Rose Odell
Woodend
Victoria

My star cat, Mia

We bought Mia from the RSPCA when she was six weeks old. Since then we have become good friends. She likes to watch television, dance – especially to a waltz – and gives me bear hugs standing on her back two feet. She is smart and can open any slide door, like the window in the bathroom.

Mia is overweight and I am trying to put her on a diet. In my previous unit, I had a latch hook on the cupboard and couldn't figure out how Mia was getting her biscuits. So one day I pretended to go out and hid behind the door.

Teamwork – I watched as both my younger cat, Brindabella, and Mia jumped up on the bench. Brindabella then got onto Mia's back and flicked the lock open. She opened the cupboard door and then opened their box of biscuits.

Mia eats her food very quickly so that she can eat Brindabella's as well. If she doesn't want to move she just rolls over. She sits next to Brindabella so that she will wash her.

Both of my cats, of course, sleep on my bed and take my blanket, even though they have their own. Mia sits on my chair as I am going to sit down. She brings me my slippers in winter in exchange for a cuddle. She has a favourite toy which looks like a sock, filled with

Smarter than Joey

stuffing, with a bell on it. My Dad and Mum laugh at her carrying it around everywhere and she only gives it to people she likes. We have to take it everywhere with us. I think my cats are very intelligent and have me well trained.

>Sharon Francis
>Tewantin
>Queensland

7

Smart animals help other animals

Her responsibility

A few years ago we had two dogs, now both gone. One, a black bitzer named Rambo, was epileptic. The other was a female sheltie called Lady.

When Rambo sensed he was having a fit he would run to us to hold him, and Lady would get most concerned and try to sit next to him.

One day my husband and I came home and the two dogs were not at the gate to greet us. Upon entering and going round the back, we found Rambo in the throws of a fit and Lady standing over him, with a paw holding him down. We were utterly amazed and thought it was so wonderful what she was trying to do for him. It was as if she knew we held him at such times and this time it was up to her.

Jan Betar
Heckenberg
New South Wales

My hero

Around 1990 I was working for the Augusta–Margaret River Shire and was also a part-time ranger. I lived at Gracetown, 20 kilometres north of Margaret River.

In those days we had local rubbish tips and Gracetown tips were trenches dug about 50 metres long, three metres deep and five metres wide. As I was the local fire control officer I used to manage the tip, mainly by burning it in the right conditions.

Come one evening, the wind and weather were right so I drove up to the tip with my dog Scuffy. The tip was fenced but the entrance was quite wide to allow vehicles to turn around. When I approached the tip, just on dark, two kangaroos were in the enclosed area and they dashed out as I approached. I stopped upwind so that I would be out of the smoke.

I let Scuffy out to have a wander around while I lit the fire. I lit upwind first so that it would burn back slowly in order to burn better. Scuffy disappeared into the smoke. I whistled her back, but she was not happy. She was jumping and barking and wanted to go back to the tip. I realised that because she was normally fairly placid, but this time she was unhappy. By this time a lot of fire and smoke were coming out of the tip. I put Scuffy in my ute but she went mental – she cried and howled. So I let her out and she raced round to the edge and was barking into the pit.

I ran round to get her. I looked into the pit and there was a little joey caught in a piece of metal shaped like a 'V'. It was jammed just below the rib cage and it couldn't move but was still alive. The fire was all around, and very close and hot. Luckily I had my big hat on to shield my face. The joey was about one and a half metres below ground level. I went over the edge, grabbed the joey and got out quick, with only a few scorched hairs on my legs and arms. Another 20 seconds and it would have been too late.

I took the joey (a little girl) back to the ute but she went limp. I thought she had died on us. I felt her ticker. It was still going so I put her on the ground out of the smoke. Scuffy had now changed her tune and appeared to be very happy, jumping all over me.

I heard a thump behind me and looked back. About 20 metres away sat mother roo. Scuffy then went and licked the joey's nose and face. Then about a minute later the joey started to move (she must have fainted). She started to struggle so I walked away from the fire with her and let her go. Both joey and mother roo hopped off into the dark.

I picked up my little hero, gave her a big hug, told her how brave she was and went home.

That's how it was.

Roy Arthur
Augusta
Western Australia

Jack the nurse!

Jack was an Appaloosa, a spotted horse heading for the meat wagon after not fitting into the 'quiet family horse' category. He was bred from a trotting mare by a spotted stallion, probably Sugarfoot, famous in the Kilmore area of Victoria.

Well, he was uncatchable, flighty and unpredictable, and he threw me over the jumps without him. He was generally a long-term project, to say the least. The only way to catch him was to entice him into the yard with a 'food trap', pretend you were feeding him and hide all signs of riding gear. Especially my boots, as they were enough to signal 'work' for him.

If he felt like it he simply jumped over the 4′6″ farm gate to escape. Well, one very hazy, extra hot summer's day in Kilmore, one of my other horses, Madeus, a huge Holsteiner-bred warm blood, was looking very strange and not well. I thought it may have been heatstroke but it turned out to be snakebite. His tummy was all tucked up and he couldn't see. Jack was leading him around by letting Madeus put his muzzle on Jack's rump as a guide! The vet came that evening, and Jack let me catch him (a miracle) and lead him and Madeus, his patient, into the yard. Jack stood by as the most attentive nurse I've ever seen.

Madeus received the anti-venom, with the vet stating he may or may not make it, but if he was alive in the morning he'd probably pull through. All night, I checked every half hour. Jack was lying close to Madeus for comfort, worrying – like me – about his mate's future. Morning dawned with Madeus crook, but alive. Jack was nuzzling the water bucket a little closer to the sick horse's mouth, so he wouldn't have to stretch for it.

For the next three weeks Jack never left Madeus' side as he recuperated and got his strength back. Jack even leaned against him for support when he first tried to stand up. So Madeus recovered. Jack instantly became as uncatchable as ever – if not more so, if this were possible – and the two horses remained best mates.

I often saw one of them pick up a big stick to entice the other to grab the other end in his mouth so they could play tug of war. If one couldn't get the stick from the other they'd canter off down the paddock together, each holding an end, quite happy to share the bloody stick!

Deslie Gillick
Margaret River
Western Australia

Mum and dad cockatiel

They had raised a fine and handsome family of young over the years and were contented grandparents.

One night there was a bit of a ruckus in the cage and the next morning mother bird was standing on the ground. Upon further investigation it became obvious that she couldn't or wouldn't fly, though she seemed otherwise unharmed. Father bird stood sentinel on the perch just above her.

A trip to the best bird vet was in order.

Mother bird had probably bumped her head during the previous night's disturbance and now had either temporary concussion or permanent and debilitating brain damage. Only time would tell. She was given a good check-up, a special meal (being famished) and sent off home to an unknown future.

Father bird was obviously relieved to see his partner return.

Mother bird stood in the bottom of her home, bewildered but seemingly comfortable. Father bird observed. He watched, perched and still. Mother bird stood on the ground.

Hours passed.

Father bird fluttered to the ground beside his partner. He walked to the water bowl and elegantly drank. Mother bird stood still. Father bird walked back to her, stood beside her and then returned to the water bowl, sipped a little more and returned to mother bird. He repeated this action countless times, interspersed with little chatters, smooches and stationary interludes. At last, mother bird understood! She took her first tentative steps towards the water bowl and a much needed drink!

Father bird immediately walked to the wire wall of the cage and began to climb, beak and feet gripping in harmonious motion. He climbed to their favourite perch, sidled along it, climbed down the other side wall and back to mother bird.

He rested there a short while, in silent commune, beside her. Then he walked to the wire, climbed effortlessly to the perch, sidled across it, down the other side and back to his lady! He repeated this for hours, again interspersed with encouraging smooches or silent commune. She watched.

That night he slept beside her on the ground.

The next day, after the mutual morning feather preening and the guided walk to the water bowl and seed pot, father bird again immediately proceeded, with enduring patience and unfailing perseverance, with the ritual climb, demonstrating to mother bird how she could best access the safety and comfort of her favourite perch.

Many hours' labour of love later he was rewarded, and oh so proud to be perched beside his mate on their favourite branch! I watched his acts of unconditional love and devotion in absolute awe, through tear-filled eyes. And learnt so much!

Mother bird never flew again, but lived a long and happy life, watched over by her mate. What a partnership! True love.

Jacky Brand
Darwin
Northern Territory

A chook's saviour

My husband Les asked me if I would cook him a chook for lunch the following day. I said okay, as he promised to catch, kill and clean it, all ready for me. Off he went that night to the fowl house with a torch to choose a suitable chook. He found a large box with no lid, so upturned the box and put the chook under it. He decided to do the job in daylight next morning.

Next morning, taking the axe, he found the box turned over and the chook gone. He was very surprised. He told me our dog Rex must have been the guilty one. He caught the chook again the next night and put it under the box, sure that the dog wouldn't release it again.

Sure enough, next morning the chook was gone and the ground around the box all scratched up. After another effort, he placed a large weight on top of the box and came up to the house smiling and saying, 'Rex will never get the chook out this time.' Sure enough, next morning another escape, and the box had a big hole dug close to the edge to let the chook out.

Finally, we had chook the following week. Les had caught the chook, killed it straight away, plucked it and cleaned it, and we had roast fowl for dinner!

Muriel Grant
Maryborough
Queensland

Surprising intelligence

When I was 12 years old I was living in a small country town called Petford in Queensland. The town mainly consisted of railway workers and people who owned cattle properties. There was a small school with approximately 20 students and a small convenience store. On the weekends I used to do odd jobs for the lady who owned the store.

One afternoon, I left the store to go and play with the store-owner's grandchildren. They had a horse called Kenny, so we all decided to grab some bread and rope and go and catch him. Terri, one of the grandchildren, called out to Kenny. As soon as he saw the bread he came straight up to her. There were a few other horses in the paddock, as well as two brumbies that were hovering in the background.

Kenny was very cunning and as soon as he had the bread he went to take off, but Terri threw a piece of lucerne string around his neck. Terri had tied a slip knot in the string, but when Kenny pulled away, he pulled so hard that the string pulled tight and started choking him. His eyes starting to roll backwards, and he began to snort and have difficulty breathing.

We did not have the strength to break or loosen the string. We all started crying and immediately ran back to the shop to get the grandmother. She was a big lady and had trouble walking at the best of times, but without even hesitating she grabbed a handful of bread and a kitchen knife and started heading down to the paddock. She didn't even stop to lock the store.

When we got down there, to our amazement, one of the brumbies was biting at the string around Kenny's neck. We couldn't believe that an animal like that would have the intelligence to know that his friend was in danger. We startled him and he took off, and Terri's grandmother released the string with a kitchen knife.

Sharon Berra
Collinsville
Queensland

Story of a sitting pig

I am now a retired pig farmer; I farmed and bred pigs for 44 years.

Some years ago one of my sows stood on a piglet's foot and injured the piglet, so I took it up to the house where it could be looked after until its foot was mended. I also had a pet duck wandering around the garden. After a few weeks the pig and the duck became mates, and in the morning at breakfast time there would be the duck, the pig, the cat and the dog all eating their food at the same time.

Well, here is the real story. The duck had laid a batch of eggs, which I didn't know about. The piglet was now around two months old, and after all the animals had eaten their breakfast the piglet would disappear. This went on for around ten days, so one day I went round very quietly and followed the piglet.

The piglet was a bit smart too, because as I was following, it would stop, listen and look around to make sure that it wasn't being followed. Eventually I saw to my amazement that the duck had laid 23 eggs (not fertile). The duck was sitting on about three quarters of the eggs (and this is the amazing bonding between animals) and the piglet was lying across the rest of the eggs.

It was staggering to watch the piglet step over the eggs. It seemed to know not to break any, which it didn't. It would very slowly and gently lie down on the eggs. This went on for ten days until the duck knew the eggs weren't any good.

Ted Hosking
Dandenong
Victoria

An unlikely caregiver

One day when I was taking blankets for needy animals to the RSPCA I spotted a magnificent looking cat named Jerry hiding under a table. He was frightened and timid. He had been dumped by a breeder who no longer wanted to breed from him. I immediately had to take him home.

From that day on he was mine. It didn't take him long to come out of himself, and once he did I saw the most wonderful personality. He followed me everywhere and, like most Siamese cats, he meowed and meowed all the time, talking to me. I understood every word and vice versa. We were like two peas in a pod.

Well, a couple of years went by and then it happened. One summer, a week before Christmas, he disappeared. It was not like him as he never left my side. I searched and called for weeks, rang every shelter and walked the streets every night. Needless to say, I had the most awful Christmas ever. I had put ads up in every shop, but nothing. I knew it was time for me to take down the ads so I did. But then one day I was at work and got a call.

'Hello, I think we have your cat.' I was stunned, I couldn't talk. I asked where she'd got my number as I had taken down all the signs. She said she saw it in one of the shops in which I hadn't taken the sign down, the laundrette. She said the cat didn't really fit my description as it looked a lot older, but it was a Siamese and it had been living under her house for the past six weeks. This matched the time that Jer had been missing.

So I left work immediately, picked Mum up and went round to the place, stunned to see that it was the house right at the back of ours. She took me to where the cat was, and I bent down and called his name. As I bent down I only called once – 'Jer' – and a quiet little meow came out, and my heart was pounding. I knew it was him.

The tears came rolling down, and all of a sudden he hopped up and came towards me so that I could reach him. He was meowing his head off. I would never have said that cats get excited, but Jerry was meowing so much I couldn't shut him up. I finally picked him up, to find that he had been hit by a car about six weeks ago. His leg was broken in many places and was dangling, and gangrene had set in. I was so appreciative of the people who had found him, but I had to go immediately to the vet.

Jerry spent three days in hospital but unfortunately had to get his front leg amputated. When he came home he was pretty much in a mess for a while, as all the nerve endings were trying to heal where his leg used to be. A few weeks later I went round to the lady's house with a present to say thank you and also to find out a bit about what happened. I was in amazement at what she was saying.

She told me that six weeks ago she saw Jerry run into her backyard with a bad limp and didn't think much of it, until a few days later she saw her cat running under the house and then running back. She thought it very odd so she followed her cat and hid.

She couldn't believe what she saw. Her cat was catching birds and taking them to Jerry (sorry to bird lovers). This went on for weeks. I said to her, when I bent down and was halfway under the house I did see a nest of feathers.

We were both amazed that her cat – which Jerry did not know, plus Jerry was in her cat's yard – was keeping him alive all this time with birds, and feathers to keep him warm! Well, I will never forget what I heard and saw that day. Cats know! I was very upset, as I kept thinking that every night he would have heard me call his name but there was no way he could have come to me or climbed the fence. He must have been very distressed.

Jerry learned to live with his disability for the next three years, until he died of kidney failure at the young age of six, which broke my heart. I know I will never again find a cat that's so loving and understanding, even up to his death. He was certainly a fighter.

Emma Betts
Baxter
Victoria

8

Smart animals show insight

She knew!

I am writing to tell you about something strange that happened several years ago. My little mongrel dog Trixie was a friendly, good-natured animal and loved people. Visitors always received a warm welcome – until something very strange happened.

I wanted some renovations done to my home and phoned a builder who had been recommended to me. He said he'd call in on his way home, about 6 pm. When he knocked on my door, Trixie went berserk. She flew against the door, barking and growling and made it very clear she was not prepared to let him in. My son had to drag her away and shut her in his room, where she whined and scratched at the door all the time I was talking to him. I was at a loss to understand her strange behaviour, as he was a polite, professional young man.

When the man was well and truly gone we let Trixie out. She came out of that room with every hair on her neck standing on edge, then searched every inch of the house and the grounds to make sure he was really gone. All that evening she lay and watched the front door. The next day she was her old self again.

This all happened on a Monday. The following Monday I read the paper and the story was on the front page. This nice young man had gone on the rampage over the weekend and shot a woman and her child, then turned the gun on himself.

How had Trixie known about a murder before it even happened? When he came to my house Trixie could not have known what the man was going to do, or why would he have bothered coming looking for more work?

I often ponder over this and can find no explanation.

Stella Greyling
Scarlborough
Western Australia

In memory of Beau

Beau, a Maltese terrier/poodle cross, was our neighbour's dog that we cared for when they went on holiday. He was always totally happy and 'our dog' until they returned. This time Max, his owner, needed heart surgery in Brisbane. From the time they left, Beau fretted; he lay around and refused to eat. We were really concerned for him.

Then Jenny, Max's wife, rang to say the operation was a success and Max was doing fine. Beau was slumped beside me, so after hanging up the phone I said, 'Beau, your dad is going to be okay.'

Immediately Beau jumped up, yelping happily, and raced round and round the room. My husband came in and couldn't believe the change in him. He stayed happy and eating well until their return.

I will always believe Beau understood what was said, and he knew that the danger he sensed for his owners was over.

Beau, who had heart problems himself, has since died.

Joan Palmblad
Nambour
Queensland

Lola the Maltese terrier

Could she sense it?

Lola is a seven-year-old Maltese terrier.

I believe she has an unexplained power that some animals have. For example, she always knew when I was coming home from work, whether it was at a scheduled time or not. But what fascinates me most is the following story.

It was a quiet evening on September 11, 2001 in Melbourne, Australia. My husband and I decided to watch a DVD movie. Lola was asleep in her basket beside us. Three quarters of the way through the movie Lola became distressed. She started barking, howling and running from one corner of the house to the other. I thought there may have been fireworks or, worse, thunder outside – but there were none.

The howling became louder, and my husband and I became very

worried because we had never seen her in this state before. We did all we could to calm her but nothing really did the trick. After about 20 minutes or so, she went back to her basket and looked very disturbed and very sad.

It was about 11.30 pm (AEST) when the movie finished. My husband, being the avid stock trader that he is, turned to the financial channel, CNBC, to check the status of the US Dow. With great fear in our eyes, we witnessed the World Trade Center in New York on fire and a second plane hitting the South Tower. It was a tragedy that would never be forgotten.

It was not until a few weeks later that I realised Lola was probably reacting to an event which happened half a world away. I was skeptical at first, so I emailed Rupert Sheldrake, the renowned author of the book *Dogs that know when their owners are coming home and other unexplained powers of animals*. I asked him whether Lola's reaction could be an unexplained power that enabled her to sense a disastrous event such as the World Trade towers' collapse. His research assistant agreed with a simple 'Yes'!

Esther Faggianelli
Moonee Ponds
Victoria

A soul in need

One hot Tuesday afternoon in February 2000, I was walking home from my friend Natasha's place. Just as I reached my front gate, my four-year-old Dobermann cross bitch Timothy Anjelika jumped the eight plus foot high fence surrounding the backyard.

At first, I thought she was just super happy to see me. But instead of jumping up on me, Tim raced past me and across the four lanes that make up Francis Street Yarraville, a very busy truck route, especially at 4 pm, peak hour.

'Oh s**?!' I yelled, in shock as my dog had her lovely black tail whacked by a passing sedan. Still standing at the gate, I dropped my bag and ran after Tim, narrowly missing being knocked down myself. I followed Tim as she ran about 500 metres up the footpath and turned in at the gate to the Westgate golf course. I kept following. We both ran across the neatly mowed green grass and, although angry golfers were screaming at her, Tim kept running.

I was running as fast as I could, but there was no way I was going to catch up to her. After running for about a kilometre, I saw we were approaching a section of Stony Creek which passes through this golf course. Tim saw it too, but did that stop her? No way! While still running, Tim half-crouched and leapt, easily clearing the seven foot wide creek. A slight stumble might deter some dogs, but not my Timothy – she kept going. I followed on, after crossing the little wooden bridge that enables golfers to access both sides of the course.

We ran and ran, ignoring the verbal abuse from many golfers. After running a speed and distance that would leave any racing greyhound for dead, Timmy leapt into Stony Creek where it first enters the golf course, right under the Williamstown Road on-ramp for the Westgate bridge.

'Okay, now what's she up to?' I wondered, as I struggled to make the distance that my dog so easily had.

Finally I reached the edge of the creek and, with all the breath I could muster, I called to Timmy. Then I noticed she was dragging something white in her mouth as she half-swam, half-waded through the murky brown and polluted water. When she reached the bank she scrambled up, dropped the plastic bag at my feet and shook herself.

'What's that, Tim?' I asked, as I crouched down and peered inside the now torn bag. What I saw made me so angry. Inside the shopping bag was a lifeless grey and white kitten. Although I was sure it was dead, I picked it up and began rubbing it with my windcheater.

Meow the little creature cried as it looked up at me with its big yellow-green eyes, before closing them again.

'Come on, Tim,' I said to my dog, my hero. She was sitting at my feet looking up at me and whining. That was very unusual for her; she hates me paying attention to any other animals.

As we made our way home, I was worried about Tim not having her leash on. But I needn't have, she heeled better than any dog ever could. Not even a dog with a utility dog title could have performed better. Ignoring the rude golfers, we finally made it home.

The first thing I did was put the kitten in my own cat's carry box, with some food, water and cat milk. The grey and white thing turned out to be a 14-week-old British shorthair, a very frightened, wet and muddy one at that.

After organising my then one-year-old Angelus (a black male Burmese/American shorthair cross), Timothy, and the other three dogs – one of which was my then three-month-old Newfoundland pup Spike – I rang Mum at work and explained the situation. She came home immediately and drove the kitten and me to the Lord Smith Animal Hospital, where the kitten stayed overnight.

The next day, Mum and I went back to the animal hospital and adopted the poor little kitten. We named her Buffy, after Buffy the Vampire Slayer, so she would go nicely with Angelus. On the TV show, Buffy and Angel dated for a while.

Three years have passed since that eventful day. Buffy's best friends are Angelus and Spike, since Buffy is only two weeks older than Spike. For all her trouble, Timothy Anjelika (now almost eight years old) was given the RSPCA's award for 'outstanding achievement'. It was presented to her at the 2000 RSPCA Million Paws Walk, by RSPCA president Dr Hugh Wirth.

One thing I am still unclear about is, how did Tim know about Buffy? The little thing was dumped a kilometre and a half away from my backyard, where Tim was. Maybe I should call in the X-Files?

Elsa Hoggard
Yarraville
Victoria

Molly the hearing dog

My hearing dog Molly, a border collie, was trained to alert me to such things as a phone ringing, a knock on the door and a smoke detector, plus other sounds. One day she went far beyond the call of duty.

I had been speaking on the phone (one with a volume control and a telecoil which picks up magnetic fields) when I heard a *psst psst* sound, similar to two live wires touching, coming over the phone. This continued during the conversation and the overhead lights dimmed twice. Turning to Molly, I asked, 'What's going on? Something is wrong.'

Molly led me into the kitchen and sat in front of the fridge, something she had not done before. Opening the fridge door I could see nothing amiss; the interior light was on and the 20-year-old motor seemed to be clunking away as usual.

Molly's refusal to leave the area was enough for me. I turned off and disconnected the fridge, and rang the local refrigeration man. His inspection revealed that two wires were touching and had become disconnected from the drip pan.

Without Molly's help, who knows what may have happened?

Hazel Grant
Maryborough
Queensland

Smarter than Joey

Goldie the dog with Ursula's daughter, Marianne

Goldie, my true friend

I saw her first at the animal shelter, and it was love at first sight. As our family had only four members here in Australia, we welcomed Goldie warmly. Goldie, in return, became a warm and affectionate friend, not only to the family but also to visitors and neighbours. She was special.

Goldie had a brown coat with white marks and a tail she proudly wagged – unlike most English fox terriers, who have their tails cut off. She liked to sing along to music and, when prompted by us, she could make high and low sounds, as well as dance to entertain us.

We loved Goldie very much, and as she grew older she needed to visit the vet occasionally. This she took in her stride, happy that we were by her side. Much later, when she was about 12 years old, she became ill with a weak heart. We fretted over her but just loved and cared for her.

On her thirteenth birthday, at about three in the morning I was awakened by something cold and wet on my face. Goldie never came into our bedroom so I knew something was wrong. I got out of bed and Goldie led me to the back door, where we sat together for a while looking up at the stars in a beautiful night sky. I told Goldie what a wonderful friend she had been and how much she meant to me. She looked at me and we both understood this meant goodbye.

In the morning Goldie had died. My husband, the children and I took Goldie and buried her at sunrise that morning in the bush near our home.

Ursula Barker
Gorokan
New South Wales

Memories from her mother

Rhani, a Great Dane, was three months old when she joined us. She was the result of a breeding programme to obtain blues. She was black, with touches of marl on her chest and paws.

The sire was a harlequin and the bitch a brindle. The breeder had searched a particular line breed and located the bitch in North Queensland. And here lies the story. When found, the bitch was tied up with wire round her neck and had been cruelly treated. A case for the RSPCA. Brought back to Melbourne and duly joined with the harlequin, she produced seven lovely puppies. Rhani was our choice.

I have always believed in putting a collar on a pup at an early age, about three to four months. This I duly did and, oh, did we get a shock. The pup screamed and went into a convulsion. I thought she was sickening for something. Removing the collar, everything went back to normal. So I left her for about three weeks, and produced the collar again. Same result. Frightening. I rang the breeder. The reply, 'Funny thing, Mrs Schmidt. Three other owners of puppies from the litter rang and said the same thing!'

At six months, one day I produced the trainer chain and lead. 'Come Rhani, let's go for a walk.' Slipping the trainer chain on the pup, you would have thought she had been trained from birth.

Conclusion: the experiences of the mother dog had been transmitted to the pups.

Makes you wonder, doesn't it?

Rhani, as she grew, became guardian of our cattle. Each day she would inspect the calves. Nudge them if they were lying down and wait for them to get up. If they did not get up she would bark to attract our attention, calling for help.

When we had retired from the farm, and Rhani was 13 years old, she was having trouble with joint aches and pains. We thought she had arthritis. Unfortunately, a diagnosis told us she had a spinal degeneration and it was incurable.

The morning she collapsed, I called our kind vet to the house and asked him to relieve her. The only thing was to have her put to sleep. The injection prepared, the vet knelt to the dog. She struggled to rise, put her head on his lap and lifted a paw.

SHE KNEW.

Phyllis Schmidt
Colac
Victoria

9

Smart animals make us wonder

Dare to be different

Every day as I clean my teeth I watch our current St Andrews Cross spider living her life on the outside of our bathroom window.

Over a period of 14 months several of these spiders have occupied this area, obviously a prime piece of real estate for them. We have spied on their mating and delighted in the resultant nursery of littlies which, despite our concerns of being overrun, have all left to seek their fortunes elsewhere. Even their mum left and, despite feeling bereft, we grabbed the chance to clean the window. Now we again find ourselves playing host to a single spider and things are back to normal – or are they?

All webs, until recently, have been the usual cross fanning out from the four groups of legs. Even the youngsters made a very creditable effort at conforming to the regulation shape. When wind threatens, the webs are thickened up, often before it arrives. Sometimes just two of the radials are thickened, but strong or gale force winds always require the strengthening of all four. As this thickening occasionally occurs when the wind doesn't get up, we suspect that they – like the Weather Bureau – are not infallible.

Our current St Andrews Cross spider is not your common or garden character, though. She does not like to conform, and displays her individuality in her web. The radials are usually of unequal length and angle, and frequently crooked. Sometimes only one or two show up,

Smarter than Joey

The spider

for she is quite frugal with her thread. Last week three radials were tiny, and the fourth was extra long and wavering as if she had an urge to make curves. Imagine being born into a body that must always produce straight lines! As a lover of curves myself, I could appreciate the temptation.

The chore of teeth cleaning has become, for me as a struggling nonconformist, a time of encouragement and inspiration as I watch our intrepid little spider expressing her individuality. She doesn't concern herself with how others see her, but just does what comes naturally to her, not her species. Vive la différence!

Wendy Willett
Russell Island
Queensland

The sea lion and the seven dolphins

The Australian sea lion (*Neophoca cinerea*) is one of the rarest seals in the world, and Australia's only endemic seal. They are found only in southern Australian waters, from the Abrolhos Islands in the west to Kangaroo Island in the east.

With the most joyous privilege of being able to visit a sea lion colony every day for two years on an island south of Perth, Western Australia, I was able to record not only sea lion behaviour, but also dolphin behaviour. Their subtle differences and similarities gave me a great respect for their intelligence.

Often found in large, dense congregations, sea lions may appear to be very social, when in fact they are very independent. After all, it does not require a coordinated team effort to pry the reef ledges for their favourite food – octopus.

Dolphins, on the other hand, act as a group and are often dependent on one another for help. At sea, a large pod will split up into numerous small pods which each have their own call signal (in fact, each individual dolphin has his or her own signal, just as we have individual names).

When a sizable school of fish has been located, all pods will be called in, and the school of fish will be surrounded by an ever-decreasing circle of dolphins blowing air, entrapping their prey in a cylindrical wall of bubbles. Once the fish are in a tightly packed ball, only one dolphin will feed at a time, taking turns, while the others keep herding.

Independent as they may be, sea lions too can be very cooperative hunters, albeit not as sophisticated as dolphins. I have also seen them circle a school of fish, entrapping them within a cylindrical wall of bubbles before feeding on them. Behaviour probably learned from dolphins, for on one occasion I did observe a sea lion cooperate with dolphins herding fish.

It was one of the younger sea lions I named Dolphin Boy from his antics of splitting from his herd to play with dolphins. By locating a pod of seven dolphins almost two kilometres away, he demonstrated to me not only his finely tuned senses after swimming directly to them, but also an ability to form an amazing cooperative interspecies relationship.

I followed on a wave ski, and as he approached the dolphins I entered the water with mask and snorkel, hoping not only he but the dolphins too would approach me. Certainly Dolphin Boy did, turning around immediately to interact with me. As we twisted and turned around each other, the dolphins circled and watched us, talking about us in their sonar clicking language. There was nothing peculiar about this behaviour.

But then Dolphin Boy left me and sank to the bottom, lying well camouflaged in the seagrass, his fur in line with the ribbon-like plant. No amount of bubbles or twisting and turning on my part seemed to arouse Dolphin Boy's attention. Such behaviour was most peculiar.

I was about to leave the water when, to my surprise, the dolphins were now closing in on me. Or so I thought. In fact, they were closing in on Dolphin Boy.

Beds of seagrass surrounding the island offer a haven for fish to hide from predators. But it offers no protection from the dolphins' penetrating sonar, which is able to detect fish almost half a metre through sand.

Three of the dolphins ploughed their noses through the seagrass towards Dolphin Boy, while four others continued circling above, emitting sonar clicks. They too were closing in, while Dolphin Boy continued to lie motionless.

The fish had apparently sought shelter under Dolphin Boy, who must have seemed like a rock to them. But that rock was about to burst. A sudden mushroom cloud of bubbles exploded from Dolphin Boy's huge lung capacity as the dolphins closed in to within a metre of him. The fish scattered from their blubbery haven – and straight into the surrounding dolphins' trap.

The dolphins could still have managed without Dolphin Boy's help. But the fact that the bubble bomb went off at the moment the dolphins were in perfect position to take maximum advantage of the fleeing fish, and that Dolphin Boy moved into position very early in the hunt, demonstrated that he had good comprehension of the dolphins' plan.

Was it instinct? Was it the dolphins' body language? Or can sea lions understand the clicking language of dolphins? Such cooperation certainly requires some sort of interspecies communication.

To date, the most optimistic estimate of the entire population of Australian sea lions is 15,000 individuals. With still much to learn about them, it is imperative that the Australian sea lions' status of protection remains. Although there does seem to be a gradual increase in population, their main cause of death is, sadly, illegal shooting.

Many crayfishermen believe sea lions are responsible for stealing crayfish from their pots, when in fact it is probably octopus that steal the crayfish and the sea lions eat the octopus. Rather than being a nuisance, they are again (probably unwittingly) cooperating with a fellow hunter.

If only we could learn to cooperate with them, as they do with dolphins, what progress that would be.

Francis Marchant
Cottesloe
Western Australia

Wilf, the gentle feral

A few years ago, I worked for an animal welfare organisation.

One day, a very small, flea-ridden and hungry kitten was rescued from the bush. Unless a miracle happened, he certainly would have been euthanased. He needed about four weeks' hand-rearing, as kittens don't normally wean themselves from mum until they're about six weeks old.

Being a 'softie', especially for tabbies, I took him home and began the four-hourly feeding routine. This included setting my alarm clock during the night. During the day, he travelled to work and back in the safety of my T-shirt.

He was a wilful kitten – hence his name, Wilf – as he thought he didn't need or want this feeding routine, he would much rather run amok!! As he grew up, I fell in love with this naughty kitten and obviously adopted him. He was sterilised, vaccinated, etc.

He developed into the most affectionate, loving cat you could ever wish for. Unfortunately, he was still a bit feral at heart – he loved to hunt and chase anything that moved.

One Sunday morning Ray, my better half, called me outside, pointing to his feet. There on the path lay a tiny creature, not much bigger than a 50 cent piece. It was a baby mouse, completely intact – strong heartbeat etc. It didn't even have its eyes open, so it must have only been a day or two old – if that.

Ever optimistic, I placed the baby into a box with tissues and a hot-water bottle! Not even an hour had gone by before there was another baby mouse placed at Ray's feet from the mouth of Wilf. By the end of the afternoon, we had four baby mice. All healthy – not a mark on them. This I found incredible, as Wilf – being a hunter – had the sharpest fangs. Not one baby mouse had any sort of injury!

Wilf had actually rescued these babies and brought them to humans. He must have killed the mother – but why didn't he eat the babies?

Did Wilf have a conscience? Did he remember his own bad start in life?

Who knows.

Not long after all this happened, my Wilf the lion-heart decided to go walkabout after one too many house moves.

I hope to God that he's okay and happy wherever he is. This story is a true one and it is dedicated to him, my sensitive little man that I miss every day. Goodbye, my angel – please be safe.

Fran Massey
Mandurah
Western Australia

Magenta

Magenta is a female spayed Himalayan sealpoint cat, now aged 18 years. We've always had an extraordinarily close and deeply bonded relationship.

In the early 1990s, my husband and I lived on a rural property some 25 kilometres south of Hobart, and Magenta, then aged six, had known no other home. My husband and I separated, and I moved to a city property on the outskirts of Hobart. I took my three cats, one of which was Magenta, with me to this new and very different environment. All three were country cats: they knew only wide open spaces, deserted unsealed roads and low population density.

They were utterly mortified by the strange, noisy landscape into which they were now thrust, and became extremely stressed. After only days, the most highly strung of the three, Scarlett, could stand it no longer and fled. She was never recovered. Fearful that the other two would follow in Scarlett's footsteps, I drove them back to my marital home at the end of the first week. I left them in the care of my husband, with whom I believed they would feel safe. I then returned to my new city home.

That very night, my husband reported that Magenta had disappeared. Frantic, I returned to the marital home and called Magenta until almost dawn, without result. Over the next few days, I produced and distributed flyers and called on neighbouring homes, in the desperate hope that someone had seen my beloved Magenta. One week went by, then two. By the beginning of the third week I had lost hope, and the grieving process began.

Then, on the twenty-first day after Magenta had disappeared from her old country home, she arrived exhausted and badly injured – in Hobart, a journey of some 25 kilometres. As I sat watching TV that evening, I could hear the persistent, distressed cries of a cat in the distance, growing rapidly closer and, as I looked round, I saw Magenta jump onto the windowsill of the room where I was seated.

Although Magenta had never seen the city house from the outside, and had been transported to and from the house by car in a covered basket, she was able to navigate a distance of 25 kilometres through towns, suburbs and heavily trafficked freeways to locate me. She was severely emaciated, with worn and bleeding paws, and coughing and sneezing blood. But for 21 days she persisted in her quest to be reunited with me, and left the only home she had ever known to seek me out.

We rushed to a vet to treat her injuries. Upon hearing Magenta's story, he remarked that in 20 years of practice he had heard several stories of cats returning to homes, but never of a cat leaving her home to find her caretaker, as Magenta had done. How did she know which direction to head in, how did she know not to give up or turn back? How had she coped with all the dangers of the journey? And how did she know when she had reached the right suburb, and the right house?

Magenta was always a remarkable personality, but this feat was extraordinary. Now aged 18, Magenta is a fragile old lady, but she retains the amazing intelligence and fighting spirit which enabled her to survive this gruelling trek so that we could be reunited. We are both very glad she made it home.

>   Amanda Meadows
>   Margate
>   Tasmania

Bird defies reason

A small bird that lives in my garden is the most motivated and industrious creature in all the world, but it behaves like a moron. Nature has produced an aberration that puts to waste hour after hour, day after day of fruitless endeavour. That's how it seemed. The bird's looks are nothing special: dark brown with a black beak and the customary beady eye.

I have a large garden, nearly all of it planted out with native trees and shrubs. In spring, it offers a bird menu second to none and the birds recognise this with gleeful song. They're fat and laid-back; even staying on the ground if they can reach a drooping grevillea flower to save flying up to a branch. They have tall eucalypts from which they survey their territory and thick shrubs in which to build their nests, from which they flush in a bad-tempered flurry whenever I take the secateurs near.

My birds carry on most of their argy-bargy at the birdbath. Who knows what it's about, but it seems important. Although they perch for a while and chatter, flip their wings about in the water and argue, they're far from being idle. But this small bird I first mentioned not only stands out as a workaholic but has made me a changed person.

The dining room wall of my house is at right angles to the living room, which boasts a door giving onto a paved area outside. The door slides open beneath a 60 millimetre wide timber pelmet housed close under the eaves. It is a dark and cosy nook, but it was never designed for the purpose to which this bird tried desperately to put it.

Early in spring, I noticed this creature flying back and forth between the garden and the house airlifting bits and pieces in its beak to the ledge above the sliding door, which, apparently, it had chosen for a home site. Anyone could see that, with such a narrow foundation, most of the hard-earned building material would slide off and fall to the ground. Which it did, piling up behind a large azalea in an out-sized pot. I went outside and told the bird 'absolutely no'. Indifferent to its wants, I swept away the sparse collection of nest-matter that still clung to the shelf, and cleared the pile of bark, fern fronds, bits of twig and leaves and some weathered garden twine that had fallen. As I did so, this bird stared at me from a distance of some three metres, fern frond in mouth, waiting. I had no sympathy whatever for its plans and went inside feeling righteous. Immediately, it started rebuilding.

I swept away the mess a second time a few days later. The bird ignored all my efforts and, like some automaton, went on collecting and trying to make the nest stay in place. I began to rethink the situation. Any creature with such a work ethic should be given a go, so I grudgingly told it 'all right, if you must'. It wasn't going to stop, anyway.

By now I was checking the bird's activities on a fairly regular basis and sometimes the construction looked more stable, but always it fell down. I started to feel quite sorry for the silly bird, which I decided must be mentally unstable. I noticed it went away to the garden for new construction material instead of dropping down for the lost pieces piling up behind the azalea. Why? It must have been a terrible bind having to look for new bits and pieces, when enough for 20 nests had already been collected and was there on the paving.

The performance of collecting, building and falling down went on again and again. What's with this crazy bird, I asked myself angrily as I hurried one day to check its activities. I had never spent so much time watching a bird in my life. Looking out the window I noticed that a patch of moss, which had proliferated where I sprayed weeds in the lawn, was seriously disturbed. While I was still annoyed with whatever animal had caused the damage, the bird flew down and took a beakful, revealing itself as the culprit. If it kept this up, I was going to have a substantial hole in the lawn, but I suddenly realised I didn't mind how much mess the bird made if only its nest would stick.

Then at last I saw that a true nest shape had emerged, and the bird was actually in it pressing around the sides and base of its tiny home with agitated actions to firm it up. Maybe the moss had helped stick the other fragments together. I murmured 'well done' and heaved a sigh of relief. The next morning the nest had fallen down.

By now I was in agonies of distress for the poor creature, not knowing if it was getting psychologically bent by its endless failures or if it was so dumb it didn't realise it was a loser. I gave a thought to nailing a piece of plywood up there to widen the ledge and tried to reach it with an extension ladder, but the angle required to clear the azalea pot was too great and I was in danger of breaking my neck. I certainly couldn't reach it from the roof. Springtime was running out, and it caused me some anguish to wonder whether it, or a mate out there, urgently needed a home where it could lay its eggs.

Eventually I knew this bird would go on doing what it was doing until some instinct told it to stop, although it could never achieve its goal. I lamented its stupidity and could no longer bear to check its daily progress.

Then, when I did look, I found the nest established and inhabited. The bird had done it. The bird had never wavered, never changed direction. But I had come a long way myself.

Judith Hollinshed
Park Orchards
Victoria

Hamish, the house spider (otherwise known as 'Pidey')

I have no love of anything with more than four legs, and an absolute aversion to anything with eight legs. I would even go so far as to relate an incident where I was running late for work and went to grab my keys from the hook, only to find my arm suddenly halting mid-reach, happily preventing me from grabbing the huge, fiendish-looking huntsman spider sitting on them.

Plan B for an arachnophobic such as myself would be to ring work and let them know I would be late. However, the phone being located about half a foot from the key hook was a problem.

I find my skin crawling even relating this ... After some exhaustive yelling at the dread beast, I was convinced my verbal taunts weren't going to have the desired effect and, content that I had given the thing due warning that it should leave, I went in search of the bug spray.

With profuse apologies I stuck my arm at the thing from about 2.5 metres away and pushed the nozzle, only to watch in dismay as a pathetic little cough of the last of the can wheezed out.

Desperate measures were needed. I can cope as long as I can stay a respectable distance away from spiders, so I resorted to the broom in the hope that I could at least get it away from the phone (because we don't leave an unattended spider of those proportions in the house or we may wake up beside it tomorrow). As fate would have it, this was an energetic spider who, I'm sure – had it been able – would have been chuckling with glee as it ran down the broomstick at mach speed towards my outstretched arm. More screaming, and spider and broom were both flung up in the air and landed on the kitchen bench, which incidentally is on the other side of the phone and key hook.

More cussing at it was apparently needed, particularly as the heebie jeebies had set in now. I almost didn't hear the knock at the door. Thankfully, several yards away from the nasty native was the front door, so, without breaking eye contact (keep in mind that was eight eyes onto two), I backed towards the front door and found my neighbour explaining she'd heard screaming.

I'm a tall, big person and my neighbour is a slight, small elf. She courageously marched straight up to the thing, armed with nothing more than an old coffee jar, stating that all the spider probably wanted was some love and affection (I think the thing wanted some sport), deftly caught it in the jar and walked out.

I went on my way to work and deleted the event from my mind as best I could. I never thought to enquire where she had deposited it – presumably in the middle of the street somewhere. Until about a week later, when I remembered my neighbour's kindness and decided to get her some flowers and a thank you card. So over I went to have a laugh over a nice cup of coffee – where I found myself eyeballing the wretched spider, dangling by its thread from the roof, across the table (at an unrespectable distance). On my neighbour's request it let itself down so that she could stroke its back with her finger, after calling 'Pidey, Pidey, come on now.'

I am certain my look of abject horror was enough to elicit her comment 'See? All he wanted was some love and affection.' (Footnote: in the event that anyone else ever steps in, I will be insisting on watching from a respectable distance where its new residence will be.) The obvious question on my lips being, if that spider was so intelligent how come it didn't understand me telling it, most emphatically, to bugger off. For myself, I suggest we all stick to bunnies, cats and dogs.

Kim Victory
Morphett Vale
South Australia

Meeting with the old one . . .

Summer 2003 came after a fierce winter where we saw minus 12 degrees at Stanthorpe, and minus six to minus eight on a regular basis for weeks. The drought was saturating. The cold sank so far to the roots that dehydration was the result – freeze-drying. Much of my garden died. Plants that live happily in very cold climates still need water to survive and there was pitifully little to spare. How come it didn't snow, with minus temperatures? Too dry! Summer had not brought the expected, hoped-for relief. The Wet – didn't.

Rain drenched our camp three nights in a row. We slept in bliss under our dome of tenting, with the gift of rain washing our statics and bad temper into the ocean. Christmas at Hervey Bay on the coast of Queensland – the first stage of our first holiday in a decade. We travelled north, and everywhere saw the sad results of climate out of place, out of time. The weather came from a time when there had not been enough water to share around the world, and, while flooding terrorised the other side of the globe, drought was doing the same in Australia.

A national park was the retreat from travelling that Gary and I sought and found. The wide expanse of rocks that had been the creek was baking in the heat. The evidence of masses of water passing was obvious in the smooth water-worn stones, the water-worn pebbles, the huge water-worn rocks. But not in the water we saw – a still, brooding pond.

One night, there was a kangaroo at the edge of the concrete near the amenities block. The tourists must have thought the creature was so cute! Apparently begging for attention as it stood, hunched at their feet, while they patted and petted and took photographs. I felt it looked ill and bewildered, not alert and perky as they usually are. The roo suffered many indignities, seemingly too tired to complain or even move away.

Day brought the truth of the roo's story. The poor thing came to our camp first thing in the morning. The animal seemed to have scows and was incontinent, dribbling and trickling shit and urine across the tarpaulin we had placed on the ground in front of the tent to keep the gravel out. Gary shooed it away, at first disgusted, and then fearful that it was disease-carrying. It was quite sluggish, and stopped often to gaze around in a dreamy, slack-jawed manner as a person will when in pain or very ill – or very old – eyes glazed and unseeing, or seeing far off into the distance, ears half-cocked, a bit ragged, no longer pointing to some real or imagined noise. The roo came back again, over to our camping gear and began sniffing at our water bottles.

I thought, 'The poor animal wants a drink.'

On the surface of my mind was what I had seen on the long drive to this place. Electric images, flashing like a child's cartoon drawn on a sheaf of papers, of animals in all attitudes of death on the road. Come to sip condensation from the cooling bitumen – drips of a terrible lure to trap the thirsty.

Half filling a bucket with water from the tap, I went to the kangaroo and offered it. I held it so that it did not have to grovel or stretch, so that it could stand upright while it quenched its thirst. The big head lowered to the cool water and it began to drink in great swallows, sounding just like my dog when she has a real thirst upon her. When it had drunk about a litre and was starting on the next one, I feared that it might make itself more sick and took it away, waited a moment and then offered it again. A few more sips were savoured and then the roo stood and closed its mouth, licking its lips delicately. It moved away.

I thought, 'OK, that's it, I did my best for the animal', though I was not really sure I had done the correct thing. Had I merely assuaged my own conscience? Soothed my guilt at having water to wash in, to drink, to stand under the shower, luxuriating, while the precious drops sank out of sight to some unimaginable, inaccessible, silent deep? I have a strong belief in non-interference in animals' patterns of behaviour. That includes feeding them, but in these dry times I was not convinced I should be withholding the life-sustaining drop when I had control over its use.

I approached the park rangers and told them what I had seen and done. I found out that what I had thought was a starving, ill and pitiful specimen was just very old, in roo years, according to the rangers – part of the family in the park. There was only one big grey in the park, and it was the only one left from several orphaned joeys. Tetanus, careless drivers and ticks had picked them off, one by one, and this old grey kangaroo had survived to a pampered old age. They laughed as they saw my confusion. I am embarrassed to say that, even though I was born and bred in Australia, I did not even know the gender of the animal! I didn't really look, and had called it a buck because it was so big, when the only other animals in the park were bright, active little wallabies to compare. The rangers boasted that they regularly gave it treats of banana peels and fruit. But they didn't give it water.

The park rangers regularly turned the sprinklers on to keep the lawns growing for the tourists. They do like to see green grass, even when that is not the true face of Australia in drought. Insulate them from the reality! The tourists of the previous night were not aware of the real needs of this roo. What they thought was friendliness was actually an attempt to get to the edge of the concrete, where the water was leaking and overflowing from the automatic urinal flushing system, draining instantly into the thirsty ground. Exquisite torture for the animal, while the tourists in their ignorance used it as a photo opportunity!

After quizzing the rangers, I was able to assess the roo more critically. The vacant gaze was that of an elderly dreamer. The slack jaw could have been doddering old age. The desultory pecking at the ground could have been a stomach full of banana peels. The bony frame was not starvation but the natural decline of a normally healthy animal into infirmity. The staggering, uncertain movements could have been the simple, crabbed steps of an almost complete, finished life. The incontinence was to be expected. The lack of grooming was a consequence of life's natural progression. Anthropomorphism had me by the throat! I was happy to accept the rangers' explanations, the insensitive pronouncements of policy makers who did not provide water for wild animals, when they would never have forgotten to leave out clean water for their dog every day.

While assembling the tools to prepare breakfast, I noticed the kangaroo I'd watered was moving around agilely – unlike before – intent on having a good feed of the grass nurtured by the sprinklers the rangers had turned on earlier. Nibbling and sniffing, it had direction. The animal was alert, with ears pricking and focusing on sounds beyond my hearing, eyes instinctively searching out possible danger even in this protected environment. Intermittently it would lie down on its side in the sun and groom and scratch its fur until it succeeded in rendering the raggedness smooth and tidy. The bones still groped from inside its coat, seeking daylight, but the roo seemed much happier in its skin.

A little time later I was sitting side by side with Gary on the plank seat, leaning our backs on the picnic table, enjoying a final cup of coffee fragrancing the air.

The roo came back. I finished my drink and placed the mug on the table.

The old grey kangaroo came to me, slowly hopping, gazing around slowly side to side, lips pursed, until she stood toe to toe with me. Her head was almost eye to eye with mine and she reached out her forepaws. I still thought of her as a wild animal and was at first a bit edgy. I thought she wanted a scratch under her ear and this I gently did. She wrapped her hands around my arms and dragged them down, her claws sharply curved and dangerous-looking. The roo then stretched her head high and arched back, with her hands wrapped around mine.

'Uh oh!' I thought, 'this is the way they do it when they teeter on their tails and rake up with their hind claws. I'm dead!' modifying this immediately to 'Bruised or bloody, at least!'

The long toenails lay on my bare feet and I was amazed at their length and potential. Her claws lightly scratched my arms, and I looked down and realised that each was as long as my fingers and curved like dried wattle seed cases. She obviously needed a manicure; and I saw that – because she was so old – she could no longer wear them down naturally, scratching for yams and roots and water as her kind have been bred by this harsh land to do.

The stretch was luxurious and complete, and I was reminded of a youthful roo playing and sparring with her mates long gone – none now her equal. My hands came to rest on my knees, after deciding that she didn't really want a scratch – she was not a dog, after all, and was not looking to be patronised.

I looked at Gary and he at me. Questions shuddering between us, sparking and sparkling, we looked back at the roo. She gazed at me, her eyes so beautifully placed that she could see ahead as well as sideways. Where her eyes had been small and tight before, now they were glistening, liquid, golden brown, with curled lashes framing them. Intelligent and wise, she placed both her hands, carefully crossed at the wrists, across my open hands (held ready in case she was unpredictable!).

My mind stopped. I was stunned!

Giggling like a girl, I turned to Gary and gasped, 'I can't believe this, she's thanking me!'

Sobering instantly, I treated the occasion with the respect I knew I would expect if I had made such an offering, and spoke to the elderly lady completely unselfconsciously.

'That's all right, you don't have to thank me! It was a pleasure!'

She gazed dreamily off to the left, then the right, showed me her tiny front teeth, pursed her mouth again and, lifting her hands and folding them, moved steadily off to nibble grass.

The sequel that gave this extraordinary experience an expanded dimension was enacted a short time later when Gary and I were saying 'goodbye' to the area.

We had packed our belongings and wandered down to the dry creek bed, to fossick at the bottom of what would ordinarily have been a raging torrent, bouncing and hopping from rock to rock, sparkling and spattering the green growing things along the banks. The drought had made this just a memory trapped in the shape of the stones, the hollows in the sand and the water-worn skins of the rocks.

The single small pond at the deepest point below the cascade held one fish, idly waiting for death or rain – beyond caring, beyond fighting for survival, not moving when we approached it. Dead creatures lying in attitudes of agony ringed the pool, tracing the water's retreat. A hopeful dragonfly flicked about, intent on furthering its species' span – impossible without the water-borne interim stage in its cycle. Leaves, crackling dry, made the only noise in this sink, imitating the light flick of water splashes but no restoring of life to the drying, dying plants along the creek banks. Heat baked the stones, making them hardened against the drought – unforgiving, desolate, unmovably sunk in rock, no longer mud.

Feeling comfortable in my skin, I felt optimistic in the knowledge that rain would, eventually, come to drench again this apparently dead river. Water would again flow, plants would again green and bear fruit, dragonflies would multiply and birds sing to the tunes set by the torrent of life.

Scrambling around among the stones, picking up this and that, everything was examined interestedly – you don't often have the opportunity to see what is at the bottom of a river, and it was exciting. Undiscovered treasures of crystals, washed out to come to rest right here where I was looking – this was the carat of my search. Sapphires, topaz, diamonds even! Of course, I didn't have any miners' tools, just a busy mind.

I bent to collect a souvenir, a small, milky-quartz, smooth stone – no value, but it would look nice in my fish bowl at home, permanently bathed in water, stroked by my fish's tail – what a life for such a stone! Beside the stone was a ring made of alternating red, yellow and black tiny glass beads. I immediately tried it on my toe – it slipped on as if it had always been there. I forgot the quartz, because it would have been just as happy in the next raging torrent getting worn down to gravel, to become sand. This ring, however, was a gift. It had been there for who knows how long. Slid off a girl's toe as she swam in the best swimming hole for miles. Thank you.

A 'thank you' from the people? I was taught a great respect for aboriginal knowledge, ways and beliefs in this country. I am the very last person to understand how things work. A drink in a dry place is a precious gift.

Thank you, Aunty, for giving me the opportunity to give.

Lyn A Higson
Stanthorpe
Queensland

Linus Pauling

My mother thought $2 was an exorbitant amount to pay for the grey baby budgie from the Victoria Market. I called him Linus after the comic strip character in 'Charlie Brown', but Dad always proudly claimed he was named after the chemist who discovered the benefits of Vitamin C, Linus Pauling. Hint: always let a reluctant parent choose the pet's name!

The cage was second-hand, bent and pink as I recall. Instinctively I put different diameter branches in as perches, and some of these rocked when the bird landed on them, giving Linus endless amusement as he rearranged the balance or discarded them from the cage completely. We only shut the cage door at night when he needed to sleep and have some privacy. The rest of the time he had complete access to the house.

Luckily the bathroom mirror would always distract him from exploring that room too thoroughly, as the window was without a screen. Linus would hang upside down off the mirror frame admiring himself and chatting happily for hours. I tamed him in the toilet room, as there were not many places where he could land comfortably, particularly as I was sitting on the toilet seat. When he was exhausted from flying round and round the room, he would eventually land and I would gently encourage him to climb first onto my outstretched finger. Then I would offload him onto my shoulder, cooing quietly to him, 'good boy!' Within a short time he would land on my shoulder and lean on my face cooing in the same tones, so I was fairly confident that he would be a 'talker'.

At that time, my mother was sewing my sister's wedding dress and Linus was very attracted to the pins and scissors, which were being used on the dining room table. He would perch precariously on the scissors blade as Mum was trying to cut, and she would cry, 'What are you doing?' She would shoo him off but he liked the company and the challenge of removing pins from the fabric, and the coloured thread tailor-tacks were really tricky to unpick. He would toss 100 pins out of the pin tin onto the table at a great rate, then, head cocked quizzedly, expect someone to put them back so that he could repeat the process, until he was unceremoniously popped back in his cage. *What's the matter?* he would call.

No one realised how well or how early Linus could talk, until our conversations were interrupted by another voice asking persistently *What are you doing?* and visitors would reply! He was encouraged to be a passionate Essendon supporter, calling *Come on the Dons!* and *Go Bombers*. We had a couple of long-haired dachshund dogs who knew better than to pursue Linus, but he would stalk their shiny black toenails, or walk over their heads, hanging downwards eyeballing them until they shook him off. If someone rang the front doorbell the dogs would bark, and so would the budgie! For years, whenever there was a similar sound on a game show on TV, Linus was programmed to bark in response, which was always of great amusement to us all. Similarly, when the phone rang, he would answer in a voice resembling my mother's melodious telephone voice, *Hello? No! Oh no! Well . . . !*

We were all devastated when Linus 'escaped' through an open window. We knew his chances of survival in suburbia were not too promising even though he was a competent flier. I put notices in the local milk bar windows. After about a week I was thrilled to have him returned, minus a tail feather and very weary. Apparently, Linus had recognised a girl wearing my school's uniform and had insisted on following her in his bow-legged bob until she picked him up. Her family were amazed how he talked and how tame he was. Coincidentally someone heard about the grey budgie and my 'lost pet' posters, so he found his way home again.

Linus was quite fond of a drink, and no sherry glass could be left unattended or he would be found sipping it appreciatively. We fancied he sounded tipsy after these binges: *What are you doing? Good boy, Bombers! Hello?* I am sure it was not good for him but it was really funny!

There was some pressure put on me to let my grandmother have Linus for company, and he acquired a smart new pergola-style cage with a fitted night cover. He set about demolishing both. He still had the run of her tidy little unit and no one minded his little poops as he was so entertaining and friendly. I could not believe that she would tolerate Linus stealing food off her plate – which, apart from being most unhygienic, was vastly alien to her sense of table manners! It was interesting that he would settle on Nanna's shoulder or head (looking down from a firm grip of mauve hair and tapping her glasses as she read or watched television: attention-seeking behaviour!) but he would not jump onto her finger, as she had very bad arthritis and apparently he could sense this. Everyone else could call him and he would fly over onto an outstretched finger.

Eventually, Nanna moved into a nursing home and Linus came home to me. Even at an advanced age he could be taught a new phrase with about an hour's tuition, so he remained very motivated to learn. It was amazing how often his comments would be appropriate and would lighten a tense moment during family 'discussions'. When I became engaged to be married we joked that the vow would be 'will you take Rosalind and her budgie?'

Fortunately he did, but my parents were really sad when Linus moved out of home. After all, he was only 13.

Ros Silverwood
Glen Iris
Victoria

10

Smart animals help us out

This child was not supposed to ride until he was old enough to manage a horse

Clever horses

In 1934 I was a student at the Midland Agriculture College and they had a shire horse. They would send him, pulling his cart (all on his own), from the farm to the dairy to pick up whey for the farm pigs.

He had to travel through two unfenced paddocks, through gateways and round the flower beds. He'd then back himself up to the ramp to have the whey loaded onto the cart. When loaded, he was sent back to the farm. He never once bumped the cart into gate posts.

Another horse I knew was stuck in deep mud by his front legs. His mate came to the rescue and dug away in the mud until the horse could free himself.

Mary Graves
Witchcliffe
Western Australia

Listen to the cat!

When my daughter was in primary school a girl gave her a kitten. I was not happy, but the kitten settled in at home and I accepted it.

Nine months later, at four one morning the cat jumped onto my bed and meowed loudly. I thought she needed to go out but knew there was a window open for that purpose so I ignored her. She meowed again. I ignored her. She sat on my chest and meowed once more.

When I didn't get up, she leaned down and gently bit my nose. I realised she was going to persevere until I got up, so I picked her up and put her outside. She stood on the porch, looking back at me and meowing. I told her she was 'on her own' and returned to bed. It was then I heard the cracking and roaring noises in the ceiling. The entire roof was on fire! We barely made it out before the entire roof caved in, and our home was gone in ten minutes.

What an intelligent cat!!

Joyce Elphick
Stanthorpe
Queensland

Better to give than to receive

When I was born, my grandparents 'gave me' a little black poodle cross called Dixie. As a child, I never really appreciated the trust and love this little dog gave our family.

We grew together – Dixie much faster than me relatively speaking – and would do everything together. On one occasion, I asked Dixie to sit and wait inside the middle circle of a netball ring at a park some kilometres from our home. After playing with my friends I returned home. At dusk, many hours later, my mother asked me 'Where is Dixie?' The guilt hit me like a sledgehammer. I stammered that she was at the park and I would go and get her. As fast as my little legs would take me, I returned to the park to find her sitting and waiting patiently for my return.

Despite this irresponsible act of childhood, Dixie obviously loved our family unconditionally. Every time we left the house we would give Dixie a small handful of Goodos (dog biscuits) to nibble on while we were gone. However, Dixie obviously felt it more rewarding to give rather than to receive, and upon our return everyone in the family – with the exception of my brother (whom she was not fond of after a couple of tail-pulling incidents) – would find a small cache of Goodos under their pillow on their beds.

On one occasion that stands out in memory, the family left the house in a hurry, not leaving any Goodos for Dixie and inadvertently closing the door that led to the bedrooms. An opened packet of Mint Slice chocolate biscuits had been left on the kitchen bench and were all gone when we got home. We all agreed that Dixie must have been put out by us neglecting to leave her with the Goodos, and therefore not letting her perform her generous act of giving. We could see where she had pushed a kitchen chair from under the bench to gain access to the chocolate biscuits, and assumed she had scoffed the lot as payback.

Hours later while sitting on the floor watching television, my father whooped with astonishment. There in the corner of the room behind the television was a very neat stack of Mint Slice chocolate biscuits standing ten high. The precision and skill to stack and balance those biscuits must have taken some effort, and so delicate was her touch that the biscuits hardly wore the evidence of any teeth marks. My father, who hated waste, ate those biscuits – but made sure Dixie got her fair share of Goodos, too!!

Julie Raverty
Echuca
Victoria

Peewee

He came at one of my lowest ebbs. I couldn't possibly have been any sadder. Three months prior I had put to sleep my 19-year-old cat who had cancer of the mouth. Two weeks later I had to say goodbye to my 11-year-old Afghan girl. She had a blood disorder. Numerous tests, trips to the vet, long hospital stays and still the result was the same – neither drugs nor my love could save her. It wrenched my heart, it felt like it was torn apart.

So here I was, digging in the vegetable patch with all my might, tears streaming down my face. I looked around and there he was. A black and white glossy-coated peewee looking intently at me with his black beady little eyes. He had his head on the side as if to say, *What's all this about?* I could have reached out to touch him.

He pounced on a worm I had dug up, gobbled it down and looked at me expectantly. Freeloader, I thought, and turned back to what I was doing – expecting him to fly off. He came nearer and nearer.

I tossed him a few more worms and he became bolder and bolder, until he was pecking at my toenails when I wasn't quick enough with his meal. I laughed in spite of myself. He was quite vocal and sang his peewee song.

The peewee hung around the garden over the next few days, arriving on the scene whenever I came out to the backyard. He seemed to enjoy my company and I started to talk to him, which he seemed to enjoy as well. He cocked his head to the side as if he understood what I was saying.

I had some mince steak defrosting and brought some out to him, tossing it close to me. Peewee seemed to have no fear at all. Maybe he knew I wouldn't hurt him. He became demanding, and over the next few days arrived at the back of the house and sang *pee wee, pee wee*, hopping around and making much noise. This continued until I came out with his mince steak. He eventually landed on my finger to take the food out of my hand. It was like something out of a Disney film – I was enchanted.

I had a girlfriend round for a coffee one morning and Peewee showed up. My girlfriend was blown away. 'How did you get him to sit on your finger like that?' she asked. I said that he just decided that he wanted to do it. He would arrive and tap on the window. If I tried to ignore him he'd hop up and down the window ledge. His antics made me laugh.

My husband decided that it was time for a new puppy – I think he wanted to see me smile again. Off we went to get a roly-poly, fluffy little black and white bearded collie boy, whom we named Macleod. Twelve weeks old and full of mischief that you wouldn't believe. So now Peewee had competition for his mince steak and neither of them liked it. Peewee would tap on the window and Macleod would race through the doggy door and it would be on! Peewee would torment him and fly just out of his reach. He'd dip and sway around the backyard, eventually landing on the roof of the pergola looking down intently at this cross, barking puppy. It was play! It was fun!

Eventually Peewee brought his mate to meet me. She would come close, but not too close. She liked the small pieces of mince steak thrown to her.

They were intent on making their nest and were gathering mud from beside the waterfall in our garden. Peewee still came for visits and free handouts, though not as often now as he had other things on his mind. I wondered if he would bring his babies down to meet me when the time came.

Then one day he didn't come. I never saw Peewee again, though I looked out for him. I hope nothing happened to him and that it was just time for him to move on. I always worried that perhaps he was too humanised and someone would do him a mischief – he was so tame and trusting.

Peewee came into my life when it was so low. A little wild bird that wanted to share something with me. Each peewee that I see I still whisper 'peewee'.

> Dellorayne Davidson
> Baulkham Hills
> New South Wales

Covering up

Her name is Jedda and she's a one-year-old cattle dog. I live about 200 metres from the beach, and surf every day.

Jedda goes for a surf with me. I leave my clothes in the sand dunes, and when it's time to come out of the surf Jedda brings my clothes down to me, including my sunglasses.

One day when I was sitting on the beach, a young lady went for a surf without her top. Jedda brought her clothes down to me, including the young lady's top. A bit too smart!

W J Dennes
Hat Head
New South Wales

Rider not needed

My father told me of this incident with his horse Anzac, around the 1920s at Bambaroo (near Ingham, North Queensland).

Each afternoon Pop used to catch Anzac in the home paddock and ride across Waterfall Creek to bring in the milkers. They consisted of several cows with calves at foot – the calves were penned for the night and the cows milked next morning.

One afternoon, as was usual, Pop went out to catch his horse, but found that Anzac was not in the home paddock. He had got out by knocking down the sliprail (gate) – so Pop set out on foot. Before he had gone very far he heard the cowbell. Shortly after, through the trees, he could see the cows – and behind, herding them along, was Anzac.

Quickly Pop hid, and watched as the cows and horse had their drink from the stream and continued up the bank, the horse nipping the rump of a straggler to keep it moving with the herd.

They continued up the track towards the cow yard, and when all were in the yard Anzac stood at the gate (to hold them in) and waited.

Muriel Johns
Urangan
Queensland

Giving

My beloved Indiana (Indi), a German shepherd, died last week after devoting the 12 and a half years of his life to my care. Indi was a surprise – he was born during my early thirties when I developed an incurable illness. He took it upon himself to be my guardian and full-time carer.

The first time I had a fall and landed on my behind without anything around me to help pull myself up, Indi was there nuzzling me. I put my arm around his neck to reassure him, and he arched his neck up and my arm dropped down. He came back at my arm again, which I placed on his neck. This time I held on – he took a few steps backwards, pulling me to my knees and then moved forward, positioning himself under my chest so that I could lean on him to get to my feet.

I was amazed at this and he was clearly very pleased with himself. He was to do this many more times over the years, and if he couldn't help he'd make me laugh. If I told him 'I can't get up yet darling, I'm too dizzy', he would promptly lie down on the floor with me. He'd put his head down between his paws and sigh heavily and moan – his version of 'Oh dear, isn't this terrible', until I would start to laugh. He really taught me how to get a positive perspective on things. On the mornings that I would be in great pain and didn't feel like getting up, he would be pushing his bed under mine and flicking the bedclothes up with his nose, and prancing about laughing just because it was morning. Needless to say, I would soon be out of bed.

Indi also helped me when I was a volunteer with an animal rescue group. We specialised in orphaned or dumped kittens. He was up with me for the three-hourly formula feeds and, as the kittens grew and were slowly weaned, they would crawl all over this gentle giant and he adored them.

I remember one occasion when he came running out of the bedroom with a kitten attached to his ear by its teeth. Indi was whimpering – 'get it off' – but knew if he shook his head he would hurt the kitten. Many kittens left this house, happy and well adjusted, for good homes – until my health deteriorated.

At times when I pushed myself too far, Indi could sense it. He would herd and nudge me away from whatever I was doing – he was always right!

One of the funniest things was when the care workers would come in the morning to assist me with showers. If Indi felt I'd done something I shouldn't have and hurt myself, he would 'tell' them all about it – it got to the point where the girls would listen to him, look at me and say 'What have you done?'

As you can see, Indi was an incredible spirit and his love and care supported me all his life. I'm afraid this account is rather disjointed but I'm missing him terribly. Indi gave nothing but love, joy and support all his life.

Alannah Grant
Ballina
New South Wales

Hunting for Aunt Ellie

Aunt Ellie had taken to her bed for the first time in years.

Many a time when there was illness in our family she had puttered and bounced up our rutted driveway in her dusty old brown VW, with her battered suitcase, birdcage and cat basket in the back. We would race out to greet her, knowing she always brought bags of lollies and a box of apples for teeth-cleaning afterwards. Her cheery smile would brighten any sickroom and our Mum would heave a sigh of relief to have someone to share the load, especially when most of us were down with something like measles, chickenpox, mumps or the flu.

Sassy the cat with a mouse for Aunt Ellie

Her puss, Sassy, seeking peace and quiet, would take over the sunny window seat in our front room, as we children were banned from playing there. When she felt sociable she would visit the patients and have a gentle game with all who could manage it, as if she realised they mustn't become excited. Mum used to worry she would catch the current illness, but Sassy didn't even get a runny nose and we could never find any spots from chicken pox. Perhaps it was her diet, for Aunt Ellie said she loved to eat pineapple, fruitcake, sweet corn, yoghurt and anything that came in vinegar, such as beetroot and olives.

Smokey, Aunt Ellie's blue and white budgie, had his cage in a prime spot in a corner of the kitchen. From there he could oversee all operations and comment vociferously upon any noisy ones. As most of our living was done in that big, homely room, he had plenty to watch and chatter about; but his favourite time was when we were washing up. Dad had to retire to the front room if he wanted to hear the news because Smokey's noise on top of our rattling, crashing and arguing drowned out the radio altogether.

The only way to keep Smokey relatively quiet was to open his cage, after making sure he couldn't fly out of the room. Then he had a grand time inspecting everything and knocking pencils and other small items off benches, shelves and the table. He was fascinated with the washing up and we often had to haul him out covered in froth and bubbles. If there was a book on the table with pictures of birds he would forget all else while he chattered to them.

Aunt Ellie said she borrowed books from the library so he could chat to the bird pictures. His vocabulary was large and he talked for hours. We tried to teach him 'Happy Christmas' during one visit, but he insisted on 'pretty Christmas, pretty pretty *pretty* Christmas'.

One year Aunt Ellie did not show up for an outbreak of measles and, as Dad knew his sister enjoyed her nursing visits, he decided to drive out and see if she was all right, for she had no telephone and lived three kilometres from her nearest neighbour.

On his arrival she didn't bustle out to greet him in her usual cheery way, and only her puss, Sassy, rushed up and made figures of eight around his legs while miaowing noisily. Then she scampered inside, stopping on the veranda and again inside the door to make sure Dad was following. When he turned into the living room, Sassy miaowed impatiently, insisting he follow her into Aunt Ellie's bedroom, which overlooked her lovely rose garden.

Dad was surprised at the unpleasant odour in a room that usually smelt of roses. Our aunt was lying asleep in bed, and beside it were several dead mice! Dad's exclamation woke her and her eyes danced with amusement at his expression of disgust.

'It's all right, Danny, they're just Sassy's way of making sure I don't starve while I'm ill. She thinks I can't hunt for myself so she's doing it for me!' Aunt Ellie went on to explain that her neighbour had been driving over to help while she was confined to her bed, but she was away until that evening and the nurse hadn't been in either.

Dad laughed and went off to get the shovel. The burial attended to, he prepared lunch, giving the clever puss some salmon and a piece of fruitcake as a treat. Later, as he drove away, he saw Sassy trotting proudly up to the house with a fresh mouse for Aunt Ellie's dinner.

Wendy Willett
Russell Island
Queensland

Buffy the hearing dog

Is my dog smarter than Jack? Definitely – my hearing dog Buffy never ceases to amaze me! On many occasions he has shown great love and intelligence.

Once when I was walking with Buffy, he pulled me off the footpath into the mud!! Just as I was about to growl, two children on rollerblades went flying past! I didn't hear them, and they couldn't stop – SMART! Even in supermarket queues he has pulled me aside when a staff member is trying to pass – he seems to assess the situation and acts accordingly, way beyond what he has been trained to do.

I rely on Buffy so much as I am profoundly deaf and had felt so isolated and insecure. Now that I know Buffy has everything under control, I can start to be part of life again.

Sometimes, I feel Buffy is too smart! He is known to touch me to alert me to sounds, so one night about 2 am I was woken with insistent paws! I was rather frightened actually, as on a previous occasion he

Smart animals help us out

Buffy the dog and Helen

found a prowler in my yard. I followed Buffy towards the back door, then he stopped and sat. When I eventually started thinking clearly, I realised he was in front of his food cupboard – he obviously wanted a midnight snack! He didn't get it, but it just shows how smart animals really are.

Buffy is a Lhaso Apso/Maltese terrier cross, absolutely adorable, and he just loves people and being out. As he is able to come everywhere with me he has a pretty good life, but it is the least I can do for such a precious companion.

One time, I was staying with a neighbour after a very traumatic experience. During the night I was awake and crying. Buffy went and woke my neighbour and brought her to me – truly amazing!

Buffy is my very best friend.

Helen Hopkins
Morphett Vale
South Australia

Cairo

My father bought me a horse when I was 15 years old. Up until this point I was riding horses that he had broken in and never had a horse to call my own. I called the horse Cairo. He was two and half years old, tall and gangly, part Arab, with a wonderful temperament. From then on, we were inseparable.

We lived on a property running cattle and sheep, so there was lots of work for Cairo and me to do. We got to know each other so well that we practically knew what the other was thinking. There were many incidents over the years that showed Cairo's amazing intelligence, but these are just a few.

It was my job each afternoon, after I got off the school bus, to saddle up Cairo and get the milking cows in. I would throw the reins over Cairo's neck and he would follow me over to the gate and through it. I would hop on and we would set off.

One afternoon, as I was closing the gate after we had walked through it, I turned around ready to hop on Cairo but he was gone. I saw him trotting up the paddock through the trees towards the cows, stirrup irons dangling. As I watched in amazement, Cairo walked around behind the cows and began to bring them home. He brought them up to the yards and I closed the gate behind them.

Cairo loved to spend time in our garden munching on the green grass. He had wandered over to the clothes line where my sister was putting clothes on the line with her two-year-old daughter. Rachel returned with another load, to see her little daughter Olivia sitting between Cairo's back legs. She didn't know quite what to do so stood still and so did Cairo, until Olivia crawled out from underneath. He was so gentle.

One day, my family and I came home to find that our expected visitors had already arrived and one had decided to have a ride on Cairo. We could see him cantering across the paddock in a haphazard fashion but clearly enjoying himself. He rode up to us, for us to see the great job he had done, very proud that he had saddled a horse – luckily, he chose Cairo. We were shocked to see that the girth was barely done up and the bridle was literally hanging over one ear.

As I write this I have a tear in my eye, as I miss my kind, intelligent and gentle friend, Cairo.

Heather Wisemantel
Gurley
New South Wales

Bradley the dog pays for some shopping

My dog is smarter than Jack!

My dog can count to ten! In fact, he can do addition, subtraction, multiplication and division. But even more amazingly, he can do square roots! He has also been known to do cubic and quartic roots for sceptical onlookers!

But my dog's most inspirational talents are life changing. He tells me 20 minutes before I'm going to have a seizure, so that I can find a safe place to sit down.

My life changed dramatically seven years ago. An accident, causing injury to my brain, left me having seizures constantly throughout the day. I was unable to leave the house.

When a friend gave me a beautiful ten-week-old puppy five years ago, my life began to change. I had a constant playmate and friend.

Several months later my dog began to behave strangely before I would have a seizure. It took me (and the vet) a few weeks to realise what he was doing.

Today, he is my seizure-alert dog and can do more than 138 commands to assist me in my everyday life. He has given me my life back, along with hope and a future, which would otherwise be impossible with daily seizures. He also brightens the lives of sick children in hospital when he visits them weekly and shows off more than 40 tricks.

His name is Bradley and he is my angel.

Belinda Simpson
Perth
Western Australia

Warm fuzzies from my ferrets

My introduction to ferrets was nine years ago, when someone rang us and asked if we wanted to buy one. I had seen them on TV and my husband said they made great pets, but that was the sum total of my experience with them.

I agreed to the guy's request and sent my husband off to collect the ferret. When my husband brought the animal back, I confess to being a little dismayed when I saw that it was a tiny and rather angry-looking albino. I had been hoping for one of those 'raccoon' looking ferrets.

We named her Mash . . . and so began my love affair with these incredibly funny, intelligent and compassionate creatures.

Mash wasn't an angry ferret – when she concentrated, her eyebrows used to knit together and that was what made her look like she was annoyed. She taught us everything we needed to know about what life was like with a whimsical weasel. She was so endearing and made us all laugh with her antics – who needed Prozac with a ferret in the family?!

I couldn't wait to get another one, and another, and another – they are the only moorish animals I have ever come across.

Within three years we were proudly owned by six ferrets, but at that time we were going through huge amounts of stress due to a business quickly going belly up. Our business was haemorrhaging uncontrollably, and most nights saw my husband and me lying awake fretting about how we were going to pay back all our creditors, pay the school fees and feed the family. The pressure was unbelievable.

A couple of weeks later, our youngest daughter came running into our bedroom sobbing because she had just had a dream that her grandmother (my mother) was going to die. I calmed her down and assured her that her grandma was going to be around for a long, long time, as she was a pretty tough old biddy!

My father had passed away only three years earlier, and my mother – although bedridden because of leg ulcers and riddled with osteoarthritis – was otherwise quite an indomitable woman. She had always told me she'd die when she was 90, and I believed her. That was ten years away, so I felt confident our daughter's dream wouldn't come true.

Imagine my consternation when, soon after that experience, my mother contracted pneumonia and was taken to hospital.

Adding that to my already all-pervading sense of doom made me feel almost suicidal. And yet, even through those incredibly dark days, I could still watch my ferrets and, although my heart was heavy, they still made me laugh with their silliness. There is a dance they do which we call the 'weasel war dance', where they bounce and spring around the room in absolute joyful abandon, which is such a heart-lifting display of carefree happiness! One ferret would start bouncing around on all fours, then crash into another, who would then take up the bounce and then the six of them would be off, dancing around the room and leaping at me, inviting me to join in. How could I not smile at such a display?

As my mother had a dog and a cat, my husband decided that he should spend every other night at her house so that the animals would have company, and so it would look lived-in. When he first started going, the cats would keep me company as normal. But after a few nights on my own, something very strange happened.

One night I was lying in bed reading when Chucky, our big male sable ferret, came up onto the bed and snuggled between the doona and the blanket against my leg. He was followed by his son CJ and Mulder, our black-eyed white. Then the girls came up onto the bed – Mash, Fidget, our silvermitt, and Friskie, our sable female. Three of them pressed against my left leg, and the other three snuggled against my right leg.

I was speechless at this sudden display of affection! Our ferrets would give us lots of kisses, they would come onto the bed and play, and Mash would occasionally come up and sleep under the doona with us when it was chilly – but never had I experienced such an obvious show of love, compassion and empathy from my gang of 'weasels', as we affectionately called them.

Every night my husband went to my mother's house to sleep, the ferrets would come onto the bed and snuggle against me. When my husband slept in our bed, they wouldn't come up. I guess they knew I was okay then.

My mother was in hospital for three months and finally succumbed to VRE after a fusion on her neck. Being very close to my parents, I felt my world had collapsed. And still my ferrets comforted me.

Thank heavens, events in our lives slowly started to normalise. There were still some bumps in the road and, while the ferrets stopped coming up en masse to cuddle, I always had one or two pressed against my leg when they felt I was stressed or unhappy. It was like they wanted to reassure me of their love and willingness to comfort me at all times, and took it in turns to show their affection!

Six years down the track, the memory of how my friends cared for me is still very strong. Regrettably, all my six have gone on to a better place and I still miss them all terribly. However, I do know we'll meet again one day, and then I will see them 'weasel war dancing' around me and giving me their ferret kisses. And I will be able to introduce them all to my parents, as well.

Boy! Heaven's going to be a busy place!

Nona Langley
Perth
Western Australia

Afterword

We hope you've enjoyed *Smarter than Joey*. It's been an exciting and entertaining book to create. Here's a bit of a story about how the smart animal books came to be.

Until late 1999 my life was a seemingly endless search for the elusive 'fulfilment'. I had this feeling that I was put on this earth to make a difference, but I had no idea how. This all left me feeling rather frustrated, lonely and unhappy. I'd always had a creative streak and loved animals. In my early years I spent many hours designing things such as horse saddles, covers and cat and dog beds. I even did a stint as a professional pet photographer.

Then I remembered something I was once told: do something for the right reasons and good things will come. So that's what I did. I set about starting Avocado Press and creating the first book, the New Zealand edition of *Smarter than Jack*. All the profit was to go to the Royal New Zealand SPCA.

Good things did come. People were thrilled to be a part of the book and many were first-time writers. Readers were enthralled and many were delighted to receive the book as a gift. The SPCA was over $43,000 better off and I received many calls, letters and emails from people with other ideas they would like to see come to life. What could be better than that?

How could I stop there! It was like I had created a living thing with Avocado Press; it seemed to have a life all of its own. I now had the responsibility of evolving it. It had to continue to benefit society by providing entertainment, warmth and something that people could feel part of. What an awesome responsibility and opportunity, albeit a bit of a scary one!

Avocado Press, as you may have guessed, is a little different. We are about more than just creating books; we're about sharing information and experiences, and developing things in innovative ways. One way we do this is by partnering with other organisations on specific projects to reach like-minded customers; for example, the SPCA. It's a model that everyone benefits from.

We feel it's possible to run a successful business that is both beneficial to customers and gives back to the community. We want people to enjoy and talk about our books; that way, ideas are shared and the better it becomes for everyone.

Future books you can be involved in

We're planning more books of smart animal stories. Your true stories are now being sought. Check out the details on the next page.

To support this book and others, we've created a *Smarter than Jack* web site with entertaining and interesting animal stories, information on upcoming books and ordering facilities. You may also like to sign up to receive our email newsletter.

Check out www.smarterthanjack.com.

Lastly, we'd love to hear your ideas on what you would like to read and write about in the future and how to make the next books even better.

Submit a smart animal story

Here's how to go about sending us a story:
1. The stories must be true and should be about 200 to 1000 words in length. They may be edited for publication.
2. Photographs and illustrations are welcome if they enhance the story, and if used will appear in black and white.
3. Submissions can be sent by email, post or via the web site www.smarterthanjack.com. Remember to include your name, postal and email address and phone number, and indicate if you do not wish your name to be included with your story.

   Email: submissions@avocadopress.com. A plain text file or email body is preferred, but MS Word formats are fine too.

   Post: Send to Avocado Press, PO Box 27003, Wellington, New Zealand. Handwritten submissions are perfectly acceptable, but if you can type them, so much the better.

4. Posted submissions will not be returned unless a stamped self-addressed envelope is provided.
5. The writers of stories selected for publication will be notified prior to publication.
6. Stories are welcome from everybody, and given the charitable nature of our projects there will be no prize money awarded, just recognition for successful submissions.
7. The SPCA and Avocado Press have the right to publish extracts from the stories received without restriction of location or publication.

Have a look at the web site for ideas, sample stories and online entry: www.smarterthanjack.com.

Your details here:

Name: _____
Address: _____
Town/city: _____
Telephone: _____
Email address: _____
Story title: _____

How to get more smart animal books

Would you like to order some more books in the *Smarter than Jack* series? Books available are *Smarter than Jack* (NZ edition), *Smarter than Jill* (NZ edition), *Smarter than Joey* (this book).

Postal orders: Avocado Press
                       PO Box 27003
                       Wellington
                       New Zealand

Telephone orders: 64-4-381 4470

Fax orders: 64-4-803 3347

Email orders: orders@avocadopress.com

Your details here:

Name: _____

Address: _____

Town/city: _____

Telephone: _____

Email address: _____

Titles requested: _____

Number of copies required: _____

Cost of copies at $19.95 each: $_____

Packaging and post per order in NZ: $ 4.00

Total: $_____

Delivery outside New Zealand and Australia: Write to or email Avocado Press to discuss payment and delivery options.

Payment options:
Cheque: please post with this order form.
Credit card: please complete the details below:

Card: Visa/MasterCard
Card number: ☐☐☐☐ ☐☐☐☐ ☐☐☐☐ ☐☐☐☐
Name on card: _____ Expiry date: ☐☐/☐☐

Thank you for your order. Please allow three weeks for delivery.

We'd like to hear your ideas

We would like to hear your ideas for future publications. What would you most like to read or write about? How could our books be improved?

Please complete the form below and post it to us at:

PO Box 27003, Wellington, New Zealand

or send us an email at ideas@avocadopress.com.

Your details here:

Name: _____
Address: _____
Town/city: _____
Telephone: _____
Email address: _____
Your suggestion: